THE INTER GALACTIC ANCIENT

SANDEEP BISHT

Copyright © Sandeep Bisht
All Rights Reserved.

ISBN 978-1-68487-288-6

This book has been published with all efforts taken to make the material error-free after the consent of the author. However, the author and the publisher do not assume and hereby disclaim any liability to any party for any loss, damage, or disruption caused by errors or omissions, whether such errors or omissions result from negligence, accident, or any other cause.

While every effort has been made to avoid any mistake or omission, this publication is being sold on the condition and understanding that neither the author nor the publishers or printers would be liable in any manner to any person by reason of any mistake or omission in this publication or for any action taken or omitted to be taken or advice rendered or accepted on the basis of this work. For any defect in printing or binding the publishers will be liable only to replace the defective copy by another copy of this work then available.

"The Inter Galactic Ancient"

Sandeep Bisht

Contents

Foreword   vii
Preface   ix
Acknowledgements   xi

HISTORY OF THE ANCIENT WORLD

CITIES, CIVILIZATIONS AND SOURCES

ACCORDING TO ANCIENT VEDIC SCRIPTURES

WHEN AND WHERE OF HINDU MYTHOLOGY

ANCIENT SCIENCES: EVERYTHING EXPLAINED

SIDDHIS AND NIDHIS

ANCIENT EGYPTIAN

PRACTICAL APPLICATION OF ANCIENT SCIENCES

INDIAN CIVILIZATION: THE VEDIC PERIOD

WHO THE GODS REALLY ARE

SECOND TYPE OF GODS WORSHIPPED IN HINDUISM

ARJUNA'S PENANCE

THE LAW OF KARMA

THE ACTUAL SECRET OF THE JOURNEY OF THE SOULS

THE GREAT EMPIRES OF ANCIENT INDIA

THE DECLINE OF EMPIRE & THE COMING OF ISLAM IN INDIA

GREEK CIVILIZATION

ROMAN CIVILIZATION

PERSIAN CIVILIZATION

BUDDHISM

JAINISM

CHINESE CIVILISATION

MOHENJO-DARO & HARAPPAN CIVILIZATION

DECLINE OF HARAPPAN CIVILIZATION

PARALLEL UNIVERSE & TIME

Contents

BODY CHAKRA

ACCORDING TO SANATAN-DHARMA (HINDU DHARMA)

THE SCIENCE OF TRAVELLING FASTER THAN THE SPEED OF LIGHT

THE SCIENCE OF CREATION OF LIVING BEINGS

ACCORDING TO SANATAN-DHARMA THE SOUL OF YOU

THE SCIENCE OF SENSE GRATIFICATION

SPIRITUALITY ACCORDING TO THE CULTURE

THINKING & GROWING RICH

STRIKING THE BALANCE BETWEEN THE INNER & OUTER ASPECT OF YOUR BODY

MOVING CLOSER TOWARDS SPIRITUAL NIRVANA

....

....

Foreword

Praise from the readers around the world.

- << Nice Book really fascinating. >> **USA**
- << I read Sandeep other book's too I seen Sandeep always write knowledgeable content that's I like. It's really good to show reality to others>> **Austria**
- << Good friends, good books, and a sleepy conscience: this is the ideal life. Nice book buddy>> **Brazil**
- << If you only read the books that everyone else is reading, you can only think what everyone else is thinking. Good book Sandeep >> **Serbia**
- << Sometimes, you read a book and it fills you with this weird evangelical zeal, and you become convinced that the shattered world will never be put back together unless and until all living humans read the book. >>**Ukraine**
- << Nice book Sandeep keep it up >> **India**
- << I find television very educating. Every time somebody turns on the set, I go into the other room and read a book. >>**Germany**

Preface

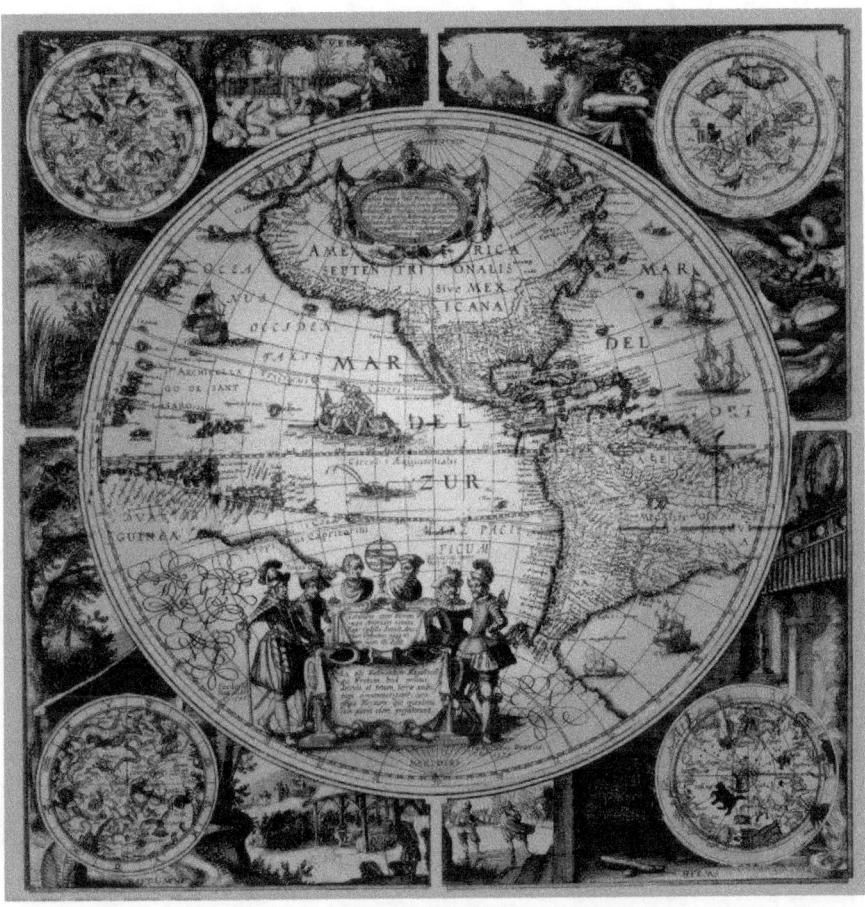

Acknowledgements

MANTRA FOR PEACE AND MEDITATIVE FOCUS.

पूर्णमदः पूर्णमिदं पूर्णात् पूर्णमुदच्यते । पूर्णस्य पूर्णमादाय पूर्णमेवावशिष्यते ॥
ॐ शान्तिः शान्तिः शान्तिः ॥

Om Purnamadah Purnamidam Purnat Purnnamudachyate
Purnasya Purnamadaya Purnamevavashishyate ||
Om Shantih Shantih Shantih |

"You are the fullness. There is fullness, here is fullness. From the fullness, the fullness is born. Remove the fullness from the fullness and the fullness alone remains."

This Mantra contains the whole secret of the mystic approach towards life. This small mantra contains the essence of the Upanishads... that nothing is insignificant; nothing is smaller than anything else. This mantra is commonly chanted as a prayer before the study of Upanishad begins. Practicing Vedic mantras allows your physical, astral and causal body to regain its natural state.

HISTORY OF THE ANCIENT WORLD

This traces the development of civilizations around the world, from the appearance of the first cities in various places around 3500–3000 B.C. until the establishment of the first true European empire under Charlemagne and the golden ages of the Abbasid Caliphate in Baghdad and the Tang dynasty in China, all during the 9th century A.D. The lectures are chronologically organized, but they interweave history with the examination of key aspects of culture, including art, literature, philosophy, religion, and architecture. We begin by looking at the earliest urban civilizations, which arose independently in Mesopotamia, Egypt, India, and China, with an emphasis on how each unique physical environment indelibly and dramatically shaped the civilization that developed in each location.

In Mesopotamia, we follow a sequence of cultures: the Sumerians, Akkadians, Babylonians, Hittites, Phoenicians, Assyrians, Chaldeans, Persians, and Sassanians. In India, we follow the growth of the Indus Valley, Vedic, and Aryan civilizations and the achievements of the Mauryan and Gupta dynasties. In China, the successive Shang, Zhou, Qin, Han, Sui, and Tang dynasties, while in the eastern Mediterranean, the preGreek Minoans and Mycenaeans are described, as is the subsequent path of classical Greek civilization, including the famed cities of Athens and Sparta and the Hellenistic world created by Alexander of Macedon. In the western Mediterranean, the fortunes of the Etruscans, Carthaginians, Romans, and various barbarian nations are all outlined. Turning to North and South America, we survey the Olmec, Chavin, Moche, Teotihuacàn, and Mayan civilizations. In Africa, the establishment of kingdoms such as Meroe, Ghana, and Axum are traced, and in Oceania, we chart the explorations of the Polynesian seafarers. Even some long-lasting hunter-gatherer societies, such as the Australian Aborigines, are examined. This comes to a close chronologically with the rise of Islam and the establishment of the Islamic Caliphates and the effect of this on Europe and the Near East.

Throughout this particular attention is given to key similarities and differences among the many civilizations studied, and so, in addition to traditionally organized lectures that provide an overview of the history and culture of a certain civilization, this course features a number of special lectures that explicitly and exclusively juxtapose illuminating aspects of widely disparate civilizations. For example, an entire lecture is devoted to comparing the epic poetry of Vedic India with Homer's Iliad. Two lectures explore the moment of intellectual questioning that occurred simultaneously in many cultures in the 6th and 5th centuries B.C. that resulted in new philosophies and religions such as Confucianism and Daoism in China, pre-Socratic philosophy in Greece, Buddhism and Jainism in India, and Zoroastrianism in Persia. A set of four interrelated lectures offers parallel biographies of five great conquerors and empire builders: Philip of Macedon and his son, Alexander the Great; Chandragupta Maurya and his grandson Asoka of India; and Shi Huangdi, the first emperor of China.

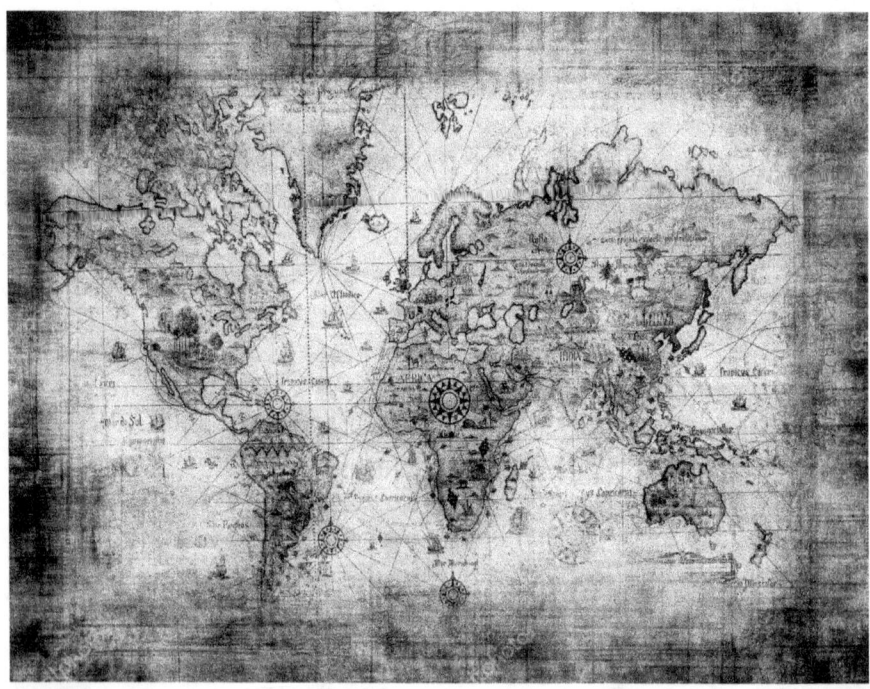

CITIES, CIVILIZATIONS AND SOURCES

This chronologically structured ancient history examines cultures around the world but also draws comparisons among different cultures as they confront similar challenges. We must keep in mind, in this or any ancient history, the limitations of studying the distant past: Much of what we study is the history of urban life—that 10 or 20 percent of the population that lived in cities.

Approaching Ancient History

- History is about people, and this course will introduce many fascinating fi gures and events. However, it will also weave these into a narrative encompassing all aspects of culture, including art, literature, architecture, philosophy, and religion.

- Comparing civilizations to each other and drawing out the similarities and differences among them invites us to consider how various historical groups sometimes made different decisions when confronted with analogous challenges.

- Let us begin with a somewhat controversial assertion: What historians traditionally call "civilization" is almost entirely an urban phenomenon: law codes, writing systems, technological innovations, art, and so forth all tended to develop in cities.

- Cities also produced individuals such as kings, emperors, inventors, philosophers, poets, artists, and warriors. Thus when we examine the history of civilization, we are studying urban history.

- The problem with this approach is that it does not represent the typical experience of the average inhabitant of the ancient world. For every person who lived in a city, about eight or nine lived out their lives on a small family farm.

A Typical Life in the Ancient World

- The typical ancient history course describes the atypical lives of a tiny minority, and we will do the same, but not before addressing, briefly, what life was like for the majority.

* Most people were born on small family farms.

* About one-quarter to one-third of babies died in their first year of life; diseases claimed many more children before puberty.

* Those who lived to adolescence had a good chance of surviving several decades of adult life, scratching out just enough food from the soil to avoid starvation. Most people died before the age of 50.

* Most people never traveled more than 20 miles from home and never saw a city. They never saw a king, took part in a battle, read a book, looked at a work of art, or heard a philosopher speak.

- This basic description applies equally well to ancient Mesopotamia, Egypt, India, or China. This sounds grim, but it was the nearly universal experience of at least 80 percent of all human beings before the Industrial Revolution.

Archaeological Evidence

- The scarcity of the surviving sources makes the study of ancient history both an exhilarating and a frustrating endeavor. Two examples illustrate the range of evidence available and the challenges the sources pose.
- In the mid-19th century, German scholar Heinrich Barth spent a number of years in Africa and discovered an astonishing archaeological site in what is now Libya: A series of 10-foot-high stone pillars, arranged in pairs, each pair topped with a lintel stone. In front of each structure was a square stone block inscribed with grooved channels.
- Barth labeled the square blocks altar stones and called the upright pillars senams. Because the structure reminded him of Stonehenge, he came to the conclusion that this was a place of worship.
- Barth's remarks inspired an Englishman named Henry Cowper to undertake a detailed study of this and similar sites in North Africa. Cowper found many such sites. The most spectacular, called Senam Semana, had no fewer than 17 trilithons.

- Cowper re-created some of the rituals performed at the altars, connecting them to ancient Babylonian gods and practices, and even suggested that the builders of Stonehenge may once have emigrated from North Africa.

- Other scholars better acquainted with Mediterranean culture soon proved, however, that the structures were actually the remains of Roman olive oil factories. The upright pillars supported the arm of the olive press, and the so-called altar stones with their grooved channels were not for lurid blood sacrifices but for directing the oil into storage containers.
- How could Cowper and Barth have been so utterly and embarrassingly wrong? Simply put, they had allowed their cultural biases and ignorance of local history and culture to cloud their judgment.

Reading Ancient Texts

- Part of Barth's and Cowper's difficulty lay in having no textual evidence to back up the physical remains. Such a situation is ripe for potential misinterpretation. However, the ancient textual evidence itself may be biased, or even worse, it may deliberately attempting to deceive us.
- Surviving documents often only give one perspective on important events. Most ancient authors intended to persuade their audience. This is not a problem if we can compare multiple accounts, but quite often only one version survives.
- Sometimes it is possible to glean more information from a text than the author meant to reveal. Consider the Behistan Inscription, a piece of propaganda that King Darius I of Persia had incised 225 feet up on the side of a sheer cliff in the Zagros Mountains.
- The opening demonstrates the flavor of the message: "I am Darius, the great king, the king of kings, the king of Persia, the king of all countries, the son of a king, the grandson of a king." The inscription lists no fewer than 23 countries he has conquered.
- It is possible to draw forth a more complex understanding of Darius's conquests from the inscription. It says of a rebellion in Armenia, "By the grace of Ahura Mazda, my army smote that rebellious army utterly." Yet in later sections, the inscription tells us that the Armenians rise up and must be defeated twice more— surely evidence of Darius's weakness,

not his strength.
- Whenever you look at historical evidence, beware of overly confident claims about ancient history; keep in mind how much of our supposed knowledge is really more speculation than firm fact. Over the course of these lectures, I will attempt to highlight points where our understanding rests on shaky ground, as well as events regarding which historians are divided.

Four Themes

- First, take note of how the physical environment in which a culture develops affects how it evolves.
- Second, keep an eye out for instances when two civilizations meet, either because of peaceful migration or militant invasion. Often, key moments of change or transformation are sparked by such interactions.
- Third, watch for innovations or experiences that seem to occur across all civilizations.
- Finally, notice how much of our contemporary culture has its origins in antiquity, so that by the end of our trip through the ancient history of the world, you should have a better sense of how the world you live in today was formed.

ACCORDING TO ANCIENT VEDIC SCRIPTURES

India is a country in South Asia whose name comes from the Indus River. The name 'Bharat' is used as a designation for the country in their constitution referencing the ancient mythological emperor, Bharata, whose story is told, in part, in the Indian epic Mahabharata.

According to the Ancient scriptures known as the Puranas (religious/historical texts written down in the 5th century CE), Bharata conquered the whole subcontinent of India and ruled the land in peace and harmony. The land was, therefore, known as Bharatavarsha (`the subcontinent of Bharata'). Hominid activity in the Indian subcontinent stretches back over 250,000 years, and it is, therefore, one of the oldest inhabited regions on the planet.

Archaeological excavations have discovered artifacts used by early humans, including stone tools, which suggest an extremely early date for human habitation and technology in the area. While the civilizations of Mesopotamia and Egypt have long been recognized for their celebrated contributions to civilization, India has often been overlooked, especially in the West, though its history and culture is just as rich. The Indus Valley Civilization (c. 7000-c. 600 BCE) was among the greatest of the ancient world, covering more territory than either Egypt or Mesopotamia, and producing an equally vibrant and progressive culture.

It is the birthplace of four great world religions - Hinduism, Jainism, Buddhism, and Sikhism - as well as the philosophical school of Charvaka which influenced the development of scientific thought and inquiry. The inventions and innovations of the people of ancient India include many aspects of modern life taken for granted today including the flush toilet, drainage and sewer systems, public pools, mathematics, veterinary science, plastic surgery, board games, yoga and meditation, as well as many more. Prehistory of India.

The areas of present-day India, Pakistan, and Nepal have provided archaeologists and scholars with the richest sites of the most ancient pedigree. The species Homo heidelbergensis (a proto-human who was an ancestor of modern Homo sapiens) inhabited the subcontinent of India centuries before humans migrated into the region known as Europe. Evidence of the existence of Homo heidelbergensis was first discovered in Germany in 1907 and, since, further discoveries have established fairly clear migration patterns of this species out of Africa.

Recognition of the antiquity of their presence in India has been largely due to the fairly late archaeological interest in the area as, unlike work in Mesopotamia and Egypt, western excavations in India did not begin in earnest until the 1920s. Though the ancient city of Harappa was known to exist as early as 1829, its archaeological significance was ignored and the later excavations corresponded to an interest in locating the probable sites referred to in the great Indian epics Mahabharata and Ramayana (both of the 5th or 4th centuries BCE) while ignoring the possibility of a much more ancient past for the region.

Archaeological excavations in the past 50 years have dramatically changed the understanding of India's past and, by extension, world history. A 4000-year-old skeleton discovered at Balathal in 2009 provides the oldest evidence of leprosy in India. Prior to this find, leprosy was considered a much younger disease thought to have been carried from Africa to India at some point and then from India to Europe by the army of Alexander the Great following his death in 323 BCE.

It is now understood that significant human activity was underway in India by the Holocene Period (10,000 years ago) and that many historical assumptions, based upon earlier work in Egypt and Mesopotamia, need to be reviewed and revised. The beginnings of the Vedic tradition in India, still practiced today, can now be dated, at least in part, to the indigenous people of ancient sites such as Balathal and their interaction and blending with the culture of Aryan migrants who arrived in the region between c. 2000-c. 1500 BCE, initiating the so-called Vedic Period (c. 1500-c.500 BCE) during which the Hindu scriptures known as the Vedas were committed to written form.

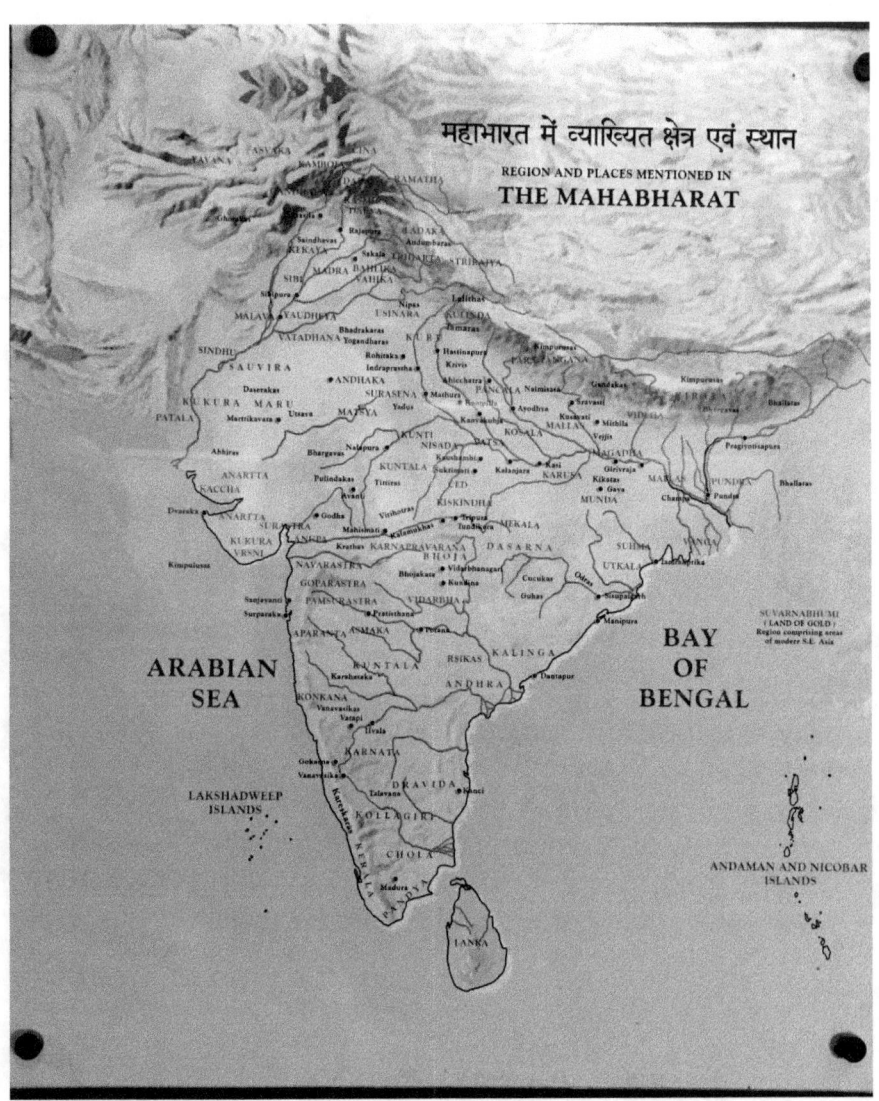

WHEN AND WHERE OF HINDU MYTHOLOGY

Hinduism is a religion that existed in Ancient Times alongside the Egyptian religion and is still existing and thriving. Hindu mythology is therefore our key to understanding the level of scientific development, lifestyle and social behaviour of the Ancient humans. Luckily for us, Hindu mythology provides this information in great details. Hindu mythology is our window into the Ancient world as it existed, as well as their understanding of the world around them. While the stories of the Ancient times are still intact in Hindu Mythology, over the years we seem to have lost the perspective in which these stories were told. In this book, one by one we will untie the knots and understand this lost perspective in order to understand our own origins, our lost sciences and our lost perspective.

For anyone who has ever wondered why our mythology gives such detailed descriptions of vimanas/ aeroplanes and nuclear bombs when according to modern science aeroplanes were invented by the Wright Brothers, and humans were hunters and gatherers at the time when these technologies were supposed to have existed or anyone who has ever been curious about what and why do we practice the rituals that we do in Hinduism, this book is an attempt to bridge the gap between mythology and modern history.

There are aspects of our modern day recorded History which the Historians themselves are unable to explain. According to History as it stands today, humans were hunters and gatherers/ nomads for a very long time and then all of a sudden two great civilizations i.e Indus Valley Civilization and Egyptian civilization, came into existence whose origins are still a mystery to all historians, this evolution from hunters and gatherers to all of a sudden developing great civilizations with advanced technology and skills is the page that is missing from present day recorded history. This page lies hundreds of feet underwater across various places around the globe and was discovered in 2002.

In 2002 Indian oceanographers discovered two cities, the size of Manhattan, off the Gulf of Khambat, lying 100 ft under water (these cities are different from the much discussed discovery of city of Dwarka discussed by popular media)

The most interesting part of this discovery is that the only time this land; where the cities have been discovered; was above the sea was before the last Ice Age. Now if we go by the description of modern history as it stands today, then during the last ice age humans were merely hunters and gatherers and could not have built any cities, especially not two cities, the size of Manhattan as have been found in the case of the submerged cities of Dwarka.

The significance of this discovery is manifold for Hindu mythology, as well as for world history and mythology, because all across the world there are many of flood myths like Noah's Arch and the myth of Manu, that describe how great floods had taken place on earth and thereafter, the survivors of the floods struggled to re-establish human civilization again on the planet. These flood myths are spread across the globe and across the cultures including Asia, West Asia, Africa etc. West Asia and Europe have the Sumerian Creation Myth, Gilgamesh Flood Myth, Genesis flood narrative of Noah's Ark, Classic Antiquity has Ancient Greek Flood Myths, Irish Flood Myths, Finnish Flood Myths. Many African cultures have an oral tradition of a flood myth including the Kwaya, Mbuti, Maasai, Mandin and Yoruba peoples, many such similar flood myths have been passed down through generations across the globe.

However, when our present day History was being written about 100-200 years ago, we did not have the advanced technology; which we have today and hence the historians had no means to check the veracity of these flood myths. These flood myths were therefore dismissed and set aside and our present day history was written on the basis of what we could see around us above the sea surface.

This is how this deep divide between present day history and mythology came into being.

However, the discovery of these cities underwater is a proof of the fact that the stories narrated in our mythology are indeed true and that there was a civilization before the present day human civilization which had the advanced technologies described in the Hindu mythology and were adept at using these technologies, some of the technologies described in our mythology are even more advanced than the technologies we are using today. But as we are well aware of the situation here in India, the site was worked upon for hardly a couple of years and hardly any work has been done on the site since 2004.

Nonetheless, having received this precious piece of information, we have found the missing perspective in which our mythological stories have been told. These major cataclysmic events across the globe are also what is called the Yuga cycle in Hinduism, yugas are actually ice-ages/ages, according to Hindu mythology there are "4 yugas" i.e 4 times when the human civilization will develop, spread and again go down in floods or natural cataclysmic events and the surviving humans will have to start civilization all over again. According to Egyptian mythology there are "5 suns" i.e 5 times the sun will hide for the longest time after the cataclysmic natural events after the ice ages/ages and human race will again have to start all over again and build a new civilization from the scratch. There are also Ancient records to say that end of each yuga was not necessarily by floods and that each time one of the 'five elements' may have played a role in ending civilization.

Our Ancient Hindu texts are actually the invaluable knowledge of the sciences, life and religious beliefs of this Ancient Civilization passed down to us by the survivors of these great floods.

So, now to give the answer of When and Where of Hindu Legends/ Mythology :-

1. WHO were these people mentioned in the Hindu Legends :-

Ans: These were the people who existed on this planet before the end of the Last Ice Age.

2. WHEN did these people mentioned in the Hindu Legends exist:-

Ans: The existence of these Ancient people on the planet ENDED some 12,000 years ago around 10,000 B.C when the Polar Ice Caps melted after thousands of years of their existence and development of their civilization. These people, however, lived on this planet for a lot longer than 12,000 years ago.

3. WHERE did these people mentioned in the Hindu Legends exist :-

Ans: They existed in the areas around India highlighted in the map below in Red colour. Remember, over here we are only discussing the period from Dwarka onwards (Dwapar Yuga). The historians are yet, to make discoveries about the Ayodhya period (Treta Yuga) of human history.

4. WHAT are these miracles and Gods discussed in Hindu Legends :-

Ans: We shall now discuss the contents of these legends given in the Hindu Mythology.

There are two aspects of the Hindu legends that we do not understand today:-

1. Miracles or the ANCIENT SCIENCES mentioned in the Ancient Hindu Texts

2. The Gods whom these Ancient people were frequently in contact with

Now discuss these Ancient Sciences in the next part.

HINDU MYTHOLOGY

WHO

WHEN

WHERE

WHAT = MIRACLES/ANCIENT SCIENCES & GODS

ANCIENT SCIENCES: EVERYTHING EXPLAINED

I bow to LG.

"The day science begins to study the non-physical phenomenon, it will make more progress in one decade than in all the previous centuries of its existence".:- Nikola Tesla

Mr Nikola Tesla is the scientist, engineer and inventor who has been credited with discovering almost the entire twenty first century, that we are living in now. Everything from Rotating magnetic field, AC motor, Tesla coil and Radio to the concepts of the mobile phone and X-Rays, he is credited with having discovered almost the entire 21st Century. Now when this genius tells us in his quote that "The day science begins to study the non-physical phenomenon, it will make more progress in one decade than in all the previous centuries of its existence".

The question that immediately crops up is, what exactly does Mr Tesla mean by the term "non-physical branch of sciences"? What are these sciences that can push the entire human civilization forward by hundreds of years.

The answer is, this non-physical branch of sciences, are the sciences that were practiced by the Ancient Civilizations that existed before the last Ice Age.

So now the question that arises for our consideration is what is this non-physical branch of sciences, how these Ancient people got to know about this non-physical branch of sciences and when did they develop it. Well, here's the news, while we did not have the advantage of modern day technology back in days when our present day history was being written, our present day technological endeavours deep into the sea are revealing remnants of Ancient civilizations that existed around the world and were flooded by sea water at the end of the last ice age when the snow caps melted. The flood myths like Noah's Ark are a result of these large scale floods that occurred at the end of the last Ice Age and the Ancient Texts that we carry with us today are the remnants of the knowledge of this Ancient Civilization passed down to us by the survivors of these catastrophic floods.

The knowledge of this ancient sub-merged civilization is common knowledge now. I would only like to add some of my observations here with respect to these sub-merged civilizations, firstly, the reason our historians seem to have conveniently ignored the 'Flood Myths' recorded in the mythology of several civilizations around the world is because back in the 18th and 19th century when our present day view of History was being written by our gramophone listening scholars, they did not have the technology to go deep into the ocean and scan the ocean floors to check the veracity of this information. However, now that we have the technology, our present day historians/scholars need to be quick in adapting to the revolutionary changes in these centuries old 'established ideas'.

Secondly, in my opinion in addition to these sub-merged civilizations there were some civilizations that were located inland at the time of these 'Great Floods' at the end of the last Ice Age. These civilizations like Indus Valley Civilization, some Native American civilizations and the Ancient Egyptian civilization with their cities located inland managed to escape the direct catastrophic effects of the Great Floods but had to deal with the drastic weather changes that accompanied the Great Floods. The rain water marks on the Sphinx in Egypt being a case in point, and therefore, these civilizations are actually a lot older than they are believed to be. The cities of these civilizations are the inland remnants of the civilizations that existed before the last ice Age.

Thirdly, I would like to point out that the best source to study the culture, sciences and beliefs of Ancient Civilizations that existed before the last Ice Age, available with us today are the Ancient Hindu texts. These Ancient Hindu texts are the knowledge database of this Ancient Civilization that was passed down to the survivors of the floods, as would be the case with any large scale catastrophic scenario; the survivors of the catastrophic event tried to preserve their knowledge for the generations to come in the form of these Ancient Hindu texts. Hinduism is the only religion that existed alongside the Egyptian religion and is still existing and thriving; we therefore don't just have the records of the Ancient Texts of Hinduism well preserved and available to us today after thousands of years, we actually have people who are practitioners of these Ancient Sciences given in these texts and can share their experiences with us. One such sub-merged city, a remnant of this Ancient Civilization that existed in India, namely the sub-merged city of Dwarka, a more than 12,000 year old city has been found off the coast of Gujrat in India, these are twin cities measuring about 7-8 kms each lying some 100 feet below sea waters.

For any civilization that existed more than 12,000 years ago to develop 7-8 km long twin cities, there has to be some form of sciences that these people were following, however, we don't see any technical equipments, any machinery that was being used by these people that can give us an idea about the level of development of their sciences.........

.....Is it possible that the sciences that these people were following was very different from the sciences that we understand today, that these Ancient sciences were non-physical in nature and that's why we can't find any physical proof or remnants of these sciences?

Now, to elaborate upon the quote by Mr Nikola Tesla regarding the non-physical branch of sciences, our scientists have recently come up with a principle called the " Holographic principle" a principle which basically states that the entire physical world around us is only a hologram, i.e. the physical world around us is not real, as against this, Ancient Hindu texts have been telling us since thousands of years that the whole world around us is just "Maya" i.e. the physical world around us is not real therefore to put it across with the help of an example both our present day sciences and our Ancient wisdom are telling us that we are living in a world that is not real and it is more like living inside a simulation or a video game. Now if we are living inside a video game then what our present day sciences (let's call them the Physical Branch of sciences as against the Ancient Nonphysical branch of sciences) are doing by breaking down the atoms into protons, neutrons, electrons and further breaking down the particles into quarks is equivalent to a character inside a video game trying to understand his reality by breaking down the pixels inside the video game.

As against this physical approach of present day sciences the Ancient people followed the non-physical branch of sciences that Mr Nikola Tesla is talking about in his quote. While our scientific knowledge of the physical world around us today comes down to breaking down the neutrons, protons and electrons into quarks, our sciences stop at this point and we have not yet managed to explain the vacuum or the empty space that exists between these particles. As against this the knowledge of these Ancient Sciences starts from this vacuum or empty space between the particles and dismisses all the atoms and molecules around this empty space as Maya or an illusion.

The Ancient Rishis/Yogis and Monks were actually practitioners/scientists who were following this non-physical branch of sciences and that is why our present day sciences are unable to explain some of the miracles performed by these Ancient Rishis and Monks even today. Now what these Ancient Rishis did was they detached themselves from the physical or the holographic world around them and went and sat on mountain tops and inside caves and they in a manner reverse engineered the human psyche/ the human soul (Consciousness) and found a way to reach back to the Maker/God of this physical world. They reverse engineered the human Consciousness the way today we see our scientists and our engineers reverse engineer a technological product. Similarly these scientists (Ancient Rishis/Yogis and Monks) reverse engineered the human being (human consciousness) to reach back to the maker and they did this with the help of yoga and hence they tried to understand the reason for the creation of this hologram and the influence that the factors within this hologram have on the human being and more importantly What is the purpose of human life in experiencing this simulation or video game.

Now to give you an idea about how advanced the knowledge of this Ancient Civilization was,

An example of the knowledge of the balance of nature in these Ancient times is given in Devistotra. A Hindu Shastra as thus :-

So long as this land,

Will have mountains,

forests and pastures That long will the Earth survive,

Sustaining you and the coming generations.

An examples of their knowledge of the origins of life in the Ocean can be found in the Atharva Veda

This earth, our mother, has

nurtured consciousness from the

slime of the primeval ocean

billions of years ago and has

sustained the human race for

countless centuries. Will we repay

our debts to our mother by

converting her into a burnt out

cinder circling the sun into

eternity?

- Atharva Veda 12.1.26, 28

To quote from Sankshipt Bhavishya Puran about the extent of Ancient people's knowledge of the Universe, a more than 5000 year old Hindu Text, Chapter One of this Sankshipt Bhavishya Puran, Brahmparva, says

Sankshipt Bhavishya Puran- Chapter 1- (5000 year old Hindu Text)

Brahmparva-"Poorva Kaal mein sara sansaar andhkaar se vyapt tha, koi padaarth drishtigat nahi hota tha"-"**In the beginning this Universe was in a state of darkness, and we could not see any element**"

"In the beginning this Universe was in a state of darkness and we could not see any element." Now don't our sciences today tell us that the Universe was in a dark dense state in the beginning? After reading these lines in a 5000 year old Hindu text, are we being rational in trusting anyone who tells us that the Ancient Civilizations had little or no knowledge about the Universe around them?

Further, there is presently a controversy among modern scientists, there is a debate about the end of the Universe about whether, (a) The Universe will keep expanding till infinity, (b) The Universe will contract after a point of expansion (c) An equilibrium will be reached and then the expansion of the Universe will stop. Now, i am going to give you some information given in the Mundaka Upanishad, this is again a thousands of year old Hindu text, now it says,

- "Angirā says, 'यथोर्णनाभिः सृजते गृह्णते च'– 'Yathornanābhihi srujate gruhnate cha' (Mundaka Upanishad: 1/1/7). Urnanābhi means a spider. O disciple! Just like a spider creates threads and makes a web, and when it desires it swallows it back, in the same way creation is made from Akshar." (Akshar here means AUM)"

Akshara or Word over here means AUM which means God (God in Sound Form) therefore, the answer to this puzzle as assessed by these Ancient Scientists in this thousands of years old text Mundaka Upanishad would be point (b) of the above debate.

Now in relation to this, i want to relate an incident from the life of Albert Einstein, though not a firm believer in God Himself Einstein did believe that there was a creator of this intelligent design of the Universe. Once; a friend of Einstein who did not believe in God and used to have frequent discussions with Einstein about it; visited Einstein. He saw a model of the solar system lying on a desk in front of Einstein. The friend asked Einstein

"Who made this model?", Albert Einstein said "No one" and then the friend asked again, "Who made this model?" Albert Einstein again replied, that no one had created the model of the solar system, then when upon asking for a third time, Albert Einstein gave the same reply, the friend got irritated and said "if no one created this model then how come it is lying on your desk". Albert Einstein's reply to this was, "You cannot believe that the small model of the solar system lying on my desk has no creator and you want me to believe that the original solar system has none!". See, this is the main difference between the Ancient Scriptures and modern Sciences, the Ancient Scriptures acknowledge the fact that there IS a creator, and explain everything from that perspective whereas modern sciences are still looking for a creator.

In addition to this, let's go back a little and consider the opinions of the people who are the founders of our modern day sciences and some of the greatest minds in our modern day Sciences and consider their opinion on these Ancient Hindu Texts.

In the 1920's quantum mechanics was created by the three great minds Heisenberg, Bohr and Schrödinger, who all read from and greatly respected the Vedas. They elaborated upon these ancient books of wisdom in their own language and with modern mathematical formulas in order to try to understand the ideas that are to be found throughout the Vedas, referred to in the ancient Sanskrit as "Brahman," "Paramatma," "Akasha" and "Atman." As Schrödinger said, "some blood transfusion from the East to the West to save Western science from spiritual anemia."

Heisenberg stated, "Quantum theory will not look ridiculous to people who have read Vedanta."

Schrodinger wrote in his book Meine Weltansicht:

"This life of yours which you are living is not merely a piece of this entire existence, but in a certain sense the whole; only this whole is not so constituted that it can be surveyed in one single glance. This, as we know, is what the Brahmins [wise men or priests in the Vedic tradition] express in that sacred, mystic formula which is yet really so simple and so clear; tat tvam asi, this is you. Or, again, in such words as "I am in the east and the west, I am above and below, I am this entire world."

"There is no kind of framework within which we can find consciousness in the plural; this is simply something we construct because of the temporal plurality of individuals, but it is a false construction... The only solution to this conflict insofar as any is available to us at all lies in the ancient wisdom of the Upanishad." (Mein Leben, Meine Weltansicht [My Life, My World View] (1961) Schrodinger, Chapter 4)

The famous Danish physicist and Nobel Prize winner, Laureate Niels Bohr (1885-1962) (pictured above), was a follower of the Vedas. He said, "I go into the Upanishads to ask questions." Both Bohr and Schrödinger, the founders of quantum physics, were avid readers of the Vedic texts and observed that their experiments in quantum physics were consistent with what they had read in the Vedas. Fritjof Capra, when interviewed by Renee Weber in the book The Holographic Paradigm (page 217–218), stated that Schrödinger, in speaking about Heisenberg, has said: "I had several discussions with Heisenberg. I lived in England then [circa 1972], and I visited him several times in Munich and showed him the whole manuscript chapter by chapter. He was very interested and very open, and he told me something that I think is not known publicly because he never published it. He said that he was well aware of these parallels. While he was working on quantum theory he went to India to lecture and was a guest of Tagore. He talked a lot with Tagore about Indian philosophy. Heisenberg told me that these talks had helped him a lot with his work in physics, because they showed him that all these new ideas in quantum physics were in fact not all that crazy. He realized there was, in fact, a whole culture that subscribed to very similar ideas. Heisenberg said that this was a great help for him. Niels Bohr had a similar experience when he went to China."

Robert Oppenheimer (1904 – 1967) learned Sanskrit in 1933 and read the Bhagavad-gita in the original, citing it later as one of the most influential books to shape his philosophy of life, stating that "The Vedas are the greatest privilege of this century."

Upon witnessing the world's first nuclear test in 1945, he instantly quoted Bhagavad-gita chapter 11, text 32, "Now I am become death, the destroyer of worlds."

Mr Nikola Tesla's association with Swami Vivekananda and his Vedanta philosophy is well known, Mr Tesla in fact was so influenced by Swami Vivekanada's philosophy that he started using Vedantic terms like Prana and Akasha to describe world around him.

Vedic texts such as the Bahgavad-gita and the Upanishads were collectively considered the most influential books ever written by eminent people like Thoreau, Kant, Schopenhauer, Schrödinger, Werner Heisenberg , Tesla, Einstein etc.

After reading these quotations from some of the greatest minds in our modern day sciences, we need to ask ourselves, are we being rational in dismissing the sciences of these Ancient Texts off hand?

A likely explanation for dismissing these Ancient texts could be that while the Sciences recognized the potential of the information provided in these texts, they dismissed them since in the light of our present day History there is no way to explain where these undeveloped Ancient People could have gathered this information from.

Well, now in the light of the recent discovery of these Ancient submerged cities, our History as well as our Sciences must stand corrected.

Now the question that arises for our consideration is, if these Ancient people knew so much about the Universe around them, what were the techniques, tools at their disposal to help them understand/research these things?

The miracles described in the Ancient texts and mythological stories are the Sciences of these Ancient People. We gather from the Ancient Hindu texts, that these Ancient people had divided their sciences into two categories, firstly the "Mayavi shaktiyan/powers" and the "Adhyatmic Shaktiyan/powers'. Mayavi Powers being powers derived from Maya. Maya has been defined in the Ancient Hindu scriptures as everything around us, everything the five senses can notice, natural as well as manmade. So the powers which we derive from natural resources, or manmade technology, those are all Mayavi Powers. The second branch of sciences, the Lost branch of Sciences i would say that exists today in these Ancient texts is the knowledge about the Adhyatmic shaktiyan or the Adhyatmic powers, as the word itself explains "Adhya" comes from the word Adhyan meaning "the study of or to study" and Atmic means Atma (soul), means the powers derived from the study of the Soul, so these are the powers derived from the soul. These are the sciences of our soul interacting with things around us. So these are the powers, the magical powers that are referred to in our mythological stories, which we do not understand today.

We come across variouus magical powers described in Ancient Hindu mythological stories like Trikaal Darshi Rishis (Rishis/Yogis who could see the past, present and future) stories about Rishis/Yogis like Maharishi Valmiki and Rishi Vashisht who are said to have had these powers of seeing the past, present and the future. There are also stories about Yogis who could go anywhere at the speed of thought, like Lord Parashurama.

So to give you a practical example, in today's terms if I was to explain these sciences, some of us in our dreams we see the future, sometimes we have premonitions about something that is going to happen to us you know. Sometimes we have telepathy about for example something good that is going to happen to our near and dear ones. We just get this intuition, this feeling that something is going to happen. Now a lot of our modern scientists have over the centuries made unsuccessful attempts to make a time machine and yet modern day sciences have not yet developed a time machine, but then how come we in our dreams move forward in time and see what is going to happen to us. Almost everyone of us has at some point in our lives had a dream where either directly or in a symbolic manner we have seen some event that is going to happen in our future. See, THESE are the powers lying dormant in our soul today. This is the lost branch of sciences that Ancient Hindu texts contain within them today. While these sciences were known to various civilizations across the world before the last ice age, the Hindu texts contain the detailed working and application these sciences and due to the benefit of an unbroken tradition of practicing these sciences, we also have yogis and rishis even today who are practicing these sciences. These are the powers of the soul, the power to go into the future and see what is going to happen. When a Rishi/Yogi tapped these powers and meditated upon it and developed it further, that is when then Rishi was called a Trikall Darshi Rishi i.e a term used for a Rishi/Yogi who could see the past, present and the future at will. These are the kind of powers that are lying dormant in our soul, that we are carrying within us everywhere, we have only lost the sciences of developing these powers.

Now let's discuss the practical yogic techniques at the disposal of the Ancient people to develop these Ancient sciences.

The way our present day sciences do everything for us, from curing diseases to reaching out into space to go to other planets similarly the Ancient yogic practices were the "YOGIC SCIENCES" of this Ancient civilization and these yogic sciences just like our present day sciences were developed to an extent where they had varied applications from improving the physical health of a person, to improving the person's psychological health and to even launching a person into space with the help of a technique popularly referred to as the technique of "Astral Body Projection". Astral Body Projection is a technique that is well documented in Ancient Hindu Texts and is a technique with the help of which a human can travel to the ends of the Universe in your astral body or cosmic body form while leaving the physical body behind on earth and therefore such human is free from the physical limitations of food, water and oxygen etc. which are the major challenges faced by humans in travelling to planets outside earth today.

In addition to this, these Ancient people also seem to have built Vimanas or Airplanes, at least 20 passages in the Rigveda (1028 hymns to the gods) refer exclusively to the flying vehicle of the Asvins. This flying machine is represented as three-storeyed, triangular and three –wheeled. It could carry at least three passengers. According to tradition the machine was made of gold, silver and iron, and had two wings. With this flying machine the Asvins saved King Bhujyu who was in distress at sea

American intelligence agency CIA has performed several documented experiments on these Ancient techniques of Astral Body Projection and remote viewing

Even today we have several Yogis in India who know and practice these techniques on a regular basis, one of them being Bapuji Dashrathbhai Patel who has the power to see the past, present and the future as well as the power to do Astral Body Projection. He explains these techniques and all the knowledge that He has gained from such travels on His YouTube channel "Bapuji Dashrathbhai Patel"

Trying to understand these Ancient Sciences without understanding the concept of the 'Soul' or Atma as its called in Hindi language, is like trying to understand our modern sciences without understanding the concept of an Atom. (Coincidently both Atom and Atma are similar sounding words). We need to ask ourselves, in not acknowledging the existence of the soul, just because our present day sciences have not yet verified the existence of the soul, which is practically the cornerstone of all the religions and civilizations of this Ancient world, are we making the same mistake that the scholars writing our present day interpretation of history made some two centuries ago in not acknowledging the existence of the pre-Ice Age civilization only because they did not have the technology to scan the ocean floors?

The description of heaven and hell given in the Ancient Texts of religions almost across the board in all religions around the world are actually the descriptions of planets where the human soul is born again and these are the planets that were visited by these Ancient people in their Astral body form and it is these planets that have been described to us as heaven and hell. Which planet the human soul will be born in again is based on the Karma that we do on this planet and that is precisely the reason why all the various religions from around the world focus on peace and love and goodness in your heart because they want your soul to be born into a better planet and not on a worse planet, because these Ancient people had understood the sciences of the soul and karma. So, the good news is we are not here to just grow old and die on this planet, the story goes on for a very

very long time. Hindu Vedas and Puranas (Ancient texts) give the descriptions of 14 Lokas (14 Star systems) where the soul is reborn in great details and explains about the life and circumstances that exist in each of the Lokas.

Moving on, while some of the Monks and Rishis living in Ashrams and Mountains and Monasteries are still aware of these sciences, two people namely Dr Steven Greer (An American doctor) and Mr Gregg Braden (An American scientist) have managed to bring out these sciences from the Ancient books and monasteries and brought these sciences out to the public, so we can sit at home comfortably and watch the youtube videos by these two great gentlemen explaining the Ancient Sciences to the public at large. Both these gentlemen are masters of two different aspects of these Ancient Sciences, though if these two work together, they can perform miracles in bringing forward these sciences.

Now, to give the readers another example about the scope of these sciences and the extent to which these sciences had been developed by this Ancient Civilization. The great scientist Dr Jagdish Chandra Bose, the Indian scientist who discovered that plants have life in them, stated that he started experimenting with plants because as a kid whenever he would hurt a plant, his mother would scold him and tell him that plants have life in them and that he should not hurt a plant. So where did his mother get this information from? She obviously got to know that from the Ancient Scriptures,

This information was a part of common knowledge in India even before it was scientifically proven.

Rishi/Sage Bhrigu explains to Bharadvaja in Chapter 184 of the Shantiparva how plants have life or in other words soul as thus :-

"the plants feel pain and pleasure and there is growth where there is a cutting. These facts prove that plants have life or Chaitanya (consciousness). It is hence that water given to the plants is taken in by them and the fire within them enables it being digested and thus stickiness and growth result."

As far as the duration of development of these sciences by the Ancient people is concerned, i would like to add that it is my strictly personal belief that as against around 5000 years of development of our present day human civilization, out of which we have only spent some 400 years developing our sciences (400 years since the apple fell on Newton's head). This Ancient civilization survived on earth for at least twice or thrice as long as our civilization and had that much more time to develop their sciences. These sciences are thus thousands of years ahead of us. To give a few examples of the advanced state of development of these Ancient sciences. What is a little know fact today, is that it is not just plants that are conscious living beings, Hindu Ancient Scriptures also tell us that the entire planet is a conscious living being, that is why it is called Dharti Mata or Mother Earth, the Sun is a conscious living being and that is why it is called Surya Devta or the Sun God and is worshipped by so many Ancient Civilizations from around the world. Infact, all the planets, stars and the entire Universe around us is a Conscious Living Being and the Hindus worship the Universe as Lord Krishna. The idea of a "living universe" is not an unknown concept. More than two thousand years ago, Plato described the universe as thus :-

"a single living creature that encompasses all living creatures within it. In this view, we live within a living system of unfathomable intelligence, subtlety, power, and patience. In turn, we appear to be evolving expressions of that living universe, infused with a knowing capacity or consciousness, and with an existence that is largely non-material in nature."- Plato"

In the summer of 2011, Dr. Matloff a major figure in what might be called the 'interstellar movement,'an Emeritus Associate Professor and Adjunct Associate Professor in the Department of Physics at New York City College of Technology as well as Hayden Associate at the American Museum of Natural History. delivered a paper in London at the British Interplanetary Society's conference on the works of philosopher and writer Olaf Stapledon, the author of Star Maker (1937). One of Stapledon's startling ideas was that stars themselves might have a form of consciousness. Greg's presentation went to work on the notion in light of anomalous stellar velocities and asked what might make such an idea possible.

The Ancient Rishis had developed methods with the help of which they could communicate with these planets and stars at the level of Consciousness and some people like Dr Steven Greer know this technique even today and in one of his Youtube videos he tells us about a dialogue he had with Mother Earth.

This is also the reason why if you jog your memory a bit, the Greek and the Roman mythology is full of stories about the planets of our solar system, like Uranus, Mars , Venus, These people knew the art of communicating with the planets at the level of consciousness. In addition to this, the Ancient Scriptures also say that our Universe is based on sound and vibrations and that is precisely the reason why Mantras are recited. A lot of attention is traditionally paid to the tonality and the sound at time of reciting the Mantras. The Vedic mantras are divided into three categories; there is a separate category of priests that are allotted to chant each type of mantra. There is a category of priests who are supposed to say a mantra in a high tone and vibration, there is a separate category of priests that are supposed to say a mantra in a medium tone and vibration and then there is a different category of priests that are supposed to say a mantra in a low tone and vibration. That is how much importance is given to the sound and the tone of the mantras and the reason for this is that the Ancient people knew that the Universe is made of sound and vibration.

As far as the vibrations of the Universe are concerned in the perspective of modern day sciences we can relate to the "String Theory". In simple words String theory would mean:.

"String theory depicts strings of energy that vibrate, but the strings are so tiny that you never perceive the vibrations directly, only their consequences. To understand these vibrations, you have to understand a classical type of wave called a standing wave — a wave that doesn't appear to be moving"

As far as the SOUND of the Universe is concerned, modern sciences are now catching up with this information as well. NASA has recently recorded the sound of Sun produces its own sound. (Recording of the sound of the Sun In fact each and every planet and star in the Universe has its own sound and the sound of all these planets and stars put together like an orchestra sounds like AUM or OM. If fact if you hear the clipping of the sound of the Sun above, this sound itself sounds like the revered Hindu symbol of AUM or OM. In fact, the importance of sound and vibration has recently again been re-established by a study which proves that the human DNA can be reprogrammed by words and frequencies.

) In fact, the names of Hindu Gods and Goddesses are said to have been specially designed for this purpose and that is why the devotees are asked to chant the names of the Gods and Goddesses.

Jai Shri Krishn

COW :Here I would also like to add, that the concept of saving Gaia or cow as taken in Hinduism to mean the animal cow, could be a misunderstanding, since the word 'Gaia' (sounds similar to Gaia/cow as pronounced in Hindi) in Greek Mythology refers to the Planet earth, so it is possible that the message conveyed by Lord Krishna was to save the planet earth/Gaia from the pollution we are causing and not the animal.

In Sanskrit itself, the term "Gau" is used for both the planet earth as well as the animal cow.

Now, the question that might come up is, how can a concept of Greek Mythology be inculcated into Hindu Mythology? The answer is simple, as explained in Chapter one, after the great cataclysmic of the last ice age the survivors of the previous civilization spread out through whichever parts of the world they could find and established civilizations in those regions again. This also explains the vast similarities in various religious practices in many ancient civilizations around the world.

SIDDHIS AND NIDHIS

Now, the question that arises for our consideration is, what are the powers that can be obtained by practicing these Ancient Sciences.

Now, to further elaborate on these Adhyatmic sciences (Sciences of the Soul), you would recall that in the Hanuman Chalisa there is a line "Asht Siddhi Nau Nidhi ke daata As Var Deen Janki Mata", means Lord Hanuman is the master of Eight Siddhis and Nine Nidhis. Now what exactly do we mean by these Siddhis and Nidhis? Now what are these Siddhis and Nidhis, i will be coming down to this, however, in the meanwhile i would like to say that the Ancients had developed these Adhyatmic powers (Powers of the soul) with two main objectives in mind, the first objective was to make human interaction on this planet to not be of the parasitic nature that it is, you know the way the human existence on this planet today is of a parasitic nature, these sciences aimed at changing that, and the second reason for which these powers were developed were to obviously gain exceptional powers which a human would not ordinarily gain.

The powers that can be obtained by these sciences are called Siddhis and Nidhis

Now I am going to go through this list of Siddhis and Nidhis, the eight Siddhis are

Anima- Reducing one's body to the size of an atom

Mahima- Increasing the size of one's body into an infinitely large size

Garima- Which means being infinitely heavy in weight

Laghima- means becoming almost weightless

Prapti- means having unrestricted access to all places, this is something that Lord Narada has, the ability to go anywhere in the Universe

Prakamya- means realizing whatever one desires

Istva- means possessing absolute lordship

Vastva- means the power to subjugate all.

Now, Five siddhis of yoga and meditation as given in the Bhagavata Purana are, the five siddhis of yoga and meditation are:

(1) trikālajñatvam: knowing the past, present and future, which is a Trikaal Darshi Rishi as i have mentioned earlier

(2) advandvam: tolerance of heat, cold and other dualities, so you see making human body independent of the climatic conditions around it. This is a part of what was mentioned earlier, a part of the Ancient Sciences which made human dependence on the planet to not be of parasitic nature. So the first thing you see is, they developed a science so that the climatic conditions around the human do not affect him

(3) para citta ādi abhijñatā: knowing the minds of others and so on

(4) agni arka ambu viṣa ādīnām pratiṣṭambhaḥ: checking the influence of fire, sun, water, poison, and so on

(5) aparajayah: remaining unconquered by others

Now ten secondary Siddhis mentioned in the Bhagavata Purana are, Ten secondary siddhis given in the Bhagavata Purana, Lord Krishna describes the ten secondary siddhis:

(1) anūrmimattvam: Being undisturbed by hunger, thirst, and other bodily appetites. So you see this is the second example of making human body independent of the resources of the planet, one is where the human body is undisturbed by the climatic conditions of the planet and the second one is where the human being is undisturbed by hunger, thirst and other bodily appetites, ok, so you can imagine that a person who does not need food, does not need water, who is unaffected by the climatic conditions around him, you can imagine that his existence on this planet would not be that dependent on the resources of the planet as a normal human being. Now next one is,

(2) dūraśravaṇa: Hearing things far away, now this Siddhi of Dursravana is a Siddhi that Mr Nikola Tesla seems to have had, it is mentioned in his autobiography "My Inventions" that he could hear things from 20 feet away and since he was a karma yogi he had developed these powers.

(3) dūradarśanam: Seeing things from far away,

(4) manojavah: Moving the body wherever thought goes. Now please pay attention Manojavaha is the power to move the body wherever the thought goes, now this power, this moving the body wherever the thought goes it menas travelling faster than the speed of light, you know, one can be sitting here and think of Mars and be there in a matter of half a second, you know, less than half a second, you know, so what we are talking about here is a speed that is faster than the speed of light which is the speed of thought. (teleportation/astral projection)

(5) kāmarūpam: Assuming any form desired

(6) parakāya praveśanam: Entering the bodies of others

(7) svachanda mṛtyuh: Dying when one desires, Bheeshma Pitamah (a mythological character from the epic Mahabharata) had this power.

(8) devānām saha krīḍā anudarśanam: Witnessing and participating in the pastimes of the Gods

(9) yathā saṅkalpa saṁsiddhiḥ: Perfect accomplishment of one's determination

(10) ājñāpratihatā gatiḥ: Orders or commands being unimpeded So these are the various powers, the various Siddhis and Nidhis, these are the major Siddhis and Nidhis which are part of the powers that were developed by these sciences of the soul.

ANCIENT EGYPTIAN

Ancient Egyptian Government: The government of ancient Egypt was a theocratic monarchy as the king ruled by a mandate from the gods, initially was seen as an intermediary between human beings and the divine, and was supposed to represent the gods' will through the laws passed and policies approved.

Ancient Egyptian Social Pyramid: The Ancient Egyptian people were grouped in a hierarchical system with the Pharaoh at the top and farmers and slaves at the bottom. Egyptian social classes had some porous borders but they were largely fixed and clearly delineated, not unlike the medieval feudal system. Clearly, the groups of people nearest the top of society were the richest and most powerful.

Ancient Egyptian Pharaoh: was believed to be a God on earth and had the most power. He was responsible for making laws and keeping order, ensuring that Egypt was not attacked or invaded by enemies and for keeping the Gods happy so that the Nile flooded and there was a good harvest.

Ancient Egyptian Vizier: was the Pharaoh's chief advisor and was sometimes also the High Priest. He was responsible for overseeing administration and all official documents had to have his seal of approval. He was also responsible for the supply of food, settling disputes between nobles and the running and protection of the Pharaoh's household.

Ancient Egyptian Nobles: ruled the regions of Egypt (Nomes). They were responsible for making local laws and keeping order in their region.

Ancient Egyptian Priests: were responsible for keeping the Gods happy. They did not preach to people but spent their time performing rituals and ceremonies to the God of their temple.

Ancient Egyptian Scribes: were the only people who could read and write and were responsible for keeping records. The ancient Egyptians recorded things such as how much food was produced at harvest time, how many soldiers were in the army, numbers of workers and the number of gifts given to the Gods.

Ancient Egyptian Soldier: were responsible for the defense of the country. Many second sons, including those of the Pharaoh often chose to join the army. Soldiers were allowed to share riches captured from enemies and were also rewarded with land for their service to the country.

Ancient Egyptian Merchants: Actually, they were more like traders carried products such as gold, papyrus made into writing paper or twisted into rope, linen cloth, and jewelry to other countries. In exchange, they brought back cedar and ebony wood, elephant tusks, panther skins, giraffe tails for fly whisks, and animals such as baboons and lions for the temples or palaces.

Ancient Egyptian Craftsmen's: Were skilled workers such as – pottery makers, leatherworkers, sculptors, painters, weavers, jewelry makers, shoemakers, tailors. Groups of craftsmen often worked together in Workshops.

Ancient Egyptian Peasants: worked the land of the Pharaoh and nobles and were given housing, food and clothes in return. Some farmers rented land from nobles and had to pay a percentage of their crop as their rent.

Ancient Egyptian Slaves: There were no slave markets or auctions in Ancient Egypt. Slaves were usually prisoners captured in war. Slaves could be found in the households of the Pharaoh and nobles, working in mines and quarries and also in temples.

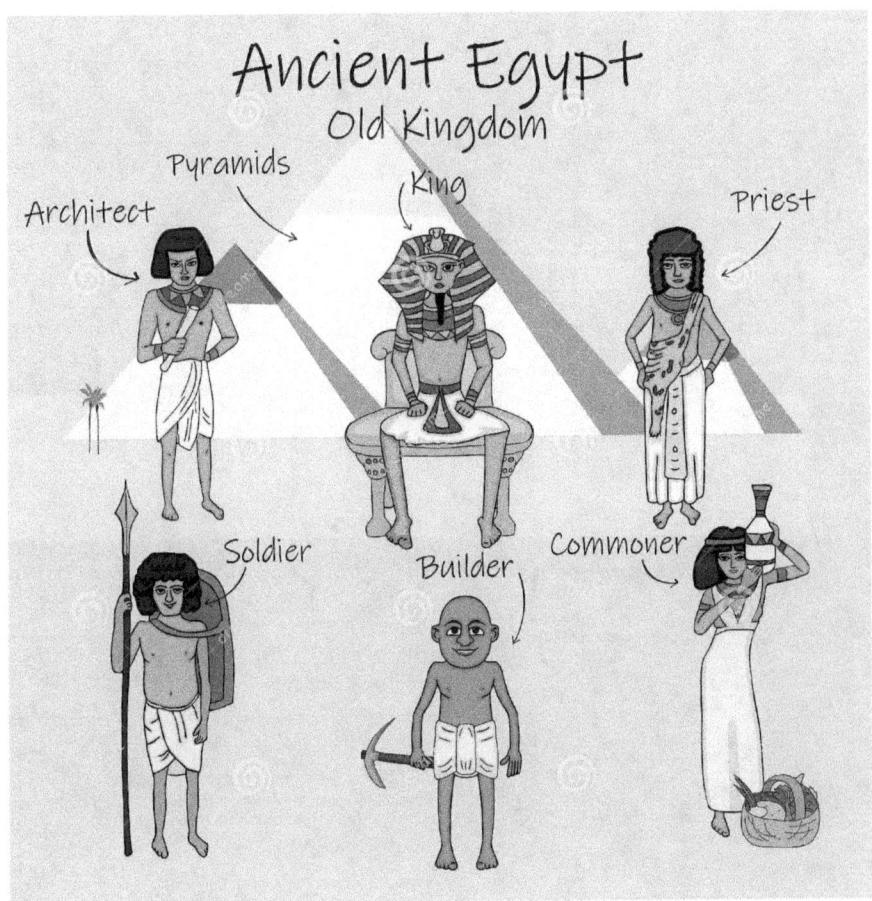

PRACTICAL APPLICATION OF ANCIENT SCIENCES

Now, if i was to discuss the practical application of this knowledge. We have already discussed that these Ancient Hindu texts are the knowledge database of entire humanity passed down to us by the survivors of the Great floods of the Last Ice Age. As already discussed, this Ancient civilization survived on earth for at least twice or thrice as long as our civilization and had that much more time to develop their sciences. These sciences are thus thousands of years ahead of us. Once this goldmine of knowledge of Ancient Sciences comes to light, we will truly understand the relevance of our History for us even today.

The most beautiful aspect of these sciences is that these sciences also have an aspect to themselves that can make a human being independent of the resources of the planet. So instead of having a parasitic relationship with the planet wherein we are dependent on the planet for everything, the human can actually be completely independent of the planet and share a nurturing bond with it as I am going to show you now. Firstly, it is quite clear that if we study these Ancient Scriptures, our sciences can take a leap forward by atleast 300 years if not more solely because of the fact that these Ancient people had access to the ends of the Universe with the help of the technique of Astral Body Projection and thus a lot more knowledge than us in the fields of physics, chemistry etc., if we study these scriptures we can even rise above the need for food and the need for woollen clothing and the need for water. These Ancient people had found out ways to make humans independent of the basic human needs of food and clothing. They had developed ways to make human body survive by simply consuming air, i,e by deriving nutrition from air itself. This knowledge is still alive and can be accessed on yogi Ray Maor's. Now you may recall that the story of Lord Krishna's parents Vasudev ji and Devaki ji who survived only on air during their tapasya (Penances), after which Lord Krishna blessed them that he would be born as their son in their next life.

Now similarly, the human body can also be nourished with the help of the Sun's rays, so we are not only dependent on food for our energy we can also derive energy from the Sun's rays, and you can access this information regarding how to nourish your body with the help of Sun's rays. If you are sceptical about this ability of the human body, you can google the name of Mr Prahlad Jani, he is a yogi who is around 70 years of age and he has not eaten anything since he was 7 years old and he has spent 10 days before the cameras 24X7 remaining before the cameras to prove this, to prove the fact that he does not eat anything and does not drink water and still survives. Similarly Mr Ray Maor, the Prana Yogi who only survives on air has also spent 7 days in front of the camera 24X7 to prove that human body can survive by taking nourishment only from air. Similarly, Mr Wim Hof, who is popularly known as the Ice Man by the media is a man who holds several records for surviving in sub-zero temperatures, and he says that he had learnt this technique of surviving in sub-zero temperatures from an Indian yogi. and he has given out the details about how to get into this practice, how to become an Ice Man in his book "Becoming An Ice Man". The only problem with the Adhyatmic Sciences or the sciences of the soul at present is that as the number of souls in the Universe increases, the powers of each individual soul keeps on decreasing, so what was easy for people to do in Ancient times with the powers of their soul, is not that easy for us to do today. We, therefore, need to use the information already available with us in these Ancient Sciences as the foundation of our modern sciences and build our sciences from there.

So you see if we go back to our scriptures our sciences can take a leap forward by atleast 300 years because the entire branch of present day sciences called Rocket Sciences can be replaced by a much simpler and more efficient sciences of Astral Body Projection, we can rise above the need for food, the need for woollen clothing, and the need for water as well.

Now over here i want to add one small thing, what is individual "Dharma", going a little away from the sciences, i want to bring forward one thing, what is individual Dharma? Now the Definition of Dharma as has been given by Dr Steven Greer in his book " Hidden Truth Forbidden Knowledge" is that every individual created being has a certain optimal point of service and an optimal role to play that is in consonance with their own unique self, so every human being has been given a few characteristics, a few personality traits, a few skills by God, which we are supposed to develop throughout our lives and it is only when we develop these skills that God has already planned for us and God has already given us that we are able to achieve an optimum combination of how to spend our life on earth, how to be productive to the best of our own ability while living on this planet and i don't just mean productive in the work sense, emotionally also, because work also somewhere affects your emotional health.

For example, if i was to give you an example, if any cricketer or Don Bradman were made to play football they might not be as good at playing football as they would be at playing cricket because that is their optimal combination, ok. Now there has been a recent study which says that even if a person spends 10,000 hours, practicing something, it is not necessary that he will become an expert at it. Now the reason for that is that even if you are putting in 10,000 hours and practicing something that you are not meant to do, you know that is not your optimal combination, then even after spending 10,000 hours you will not succeed in perfecting that art, because God has already given us inbuilt set of skills that we are supposed to develop.

INDIAN CIVILIZATION: THE VEDIC PERIOD

Whatever the reason for the abandonment of the cities, the period that followed the decline of the Indus Valley Civilization is known as the Vedic Period, characterized by a pastoral lifestyle and adherence to the religious texts known as The Vedas. Society became divided into four classes (the Varnas) popularly known as `the caste system` which were comprised of the Brahmana at the top (priests and scholars), the Kshatriya next (the warriors), the Vaishya (farmers and merchants), and the Shudra (laborers). The lowest caste was the Dalits, the untouchables, who handled meat and waste, though there is some debate over whether this class existed in antiquity.

At first, it seems this caste system was merely a reflection of one's occupation but, in time, it became more rigidly interpreted to be determined by one's birth and one was not allowed to change castes nor to marry into a caste other than one's own. This understanding was a reflection of the belief in an eternal order to human life dictated by a supreme deity.

SANATAN DHARMA HOLDS THERE IS ONE GOD, BRAHMA, WHO, CANNOT BE FULLY APPREHENDED SAVE THROUGH THE MANY ASPECTS WHICH ARE REVEALED AS THE DIFFERENT GODS OF THE HINDU PANTHEON.

While the religious beliefs which characterized the Vedic Period are considered much older, it was during this time that they became systematized as the religion of Sanatan Dharma (`Eternal Order`) known today as Hinduism (this name deriving from the Indus (or Sindus) River where worshippers were known to gather, hence, `Sindus`, and then `Hindus`). The underlying tenet of Sanatan Dharma is that there is an order and a purpose to the universe and human life and, by accepting this order and living in accordance with it, one will experience life as it is meant to be properly lived.

While Sanatan Dharma is considered by many a polytheistic religion consisting of many gods, it is actually monotheistic in that it holds there is one god, Brahman (the Self but also the Universe and creator of the observable universe), who, because of his greatness, cannot be fully apprehended save through the many aspects which are revealed as the different gods of the Hindu pantheon.

It is Brahman who decrees the eternal order and maintains the universe through it. This belief in an order to the universe reflects the stability of the society in which it grew and flourished as, during the Vedic Period, governments became centralized and social customs integrated fully into daily life across the region. Besides The Vedas, the great religious and literary works of the Puranas, the Mahabharata, Bhagavad-Gita, and the Ramayana all come from this period.

In the 6th century BCE, the religious reformers Vardhamana Mahavira (l. c. 599-527 BCE) and Siddhartha Gautama (l. c. 563-c. 483 BCE) developed their own belief systems and broke away from mainstream Sanatan Dharma to eventually create their own religions of Jainism and Buddhism, respectively. These changes in religion were a part of a wider pattern of social and cultural upheaval which resulted in the formation of city-states and the rise of powerful kingdoms (such as the Magadha Kingdom under the ruler Bimbisara) and the proliferation of philosophical schools of thought which challenged orthodox Hinduism.

Mahavira rejected the Vedas and placed the responsibility for salvation and enlightenment directly on the individual and the Buddha would later do the same. The philosophical school of Charvaka rejected all supernatural elements of religious belief and maintained that only the senses could be trusted to apprehend the truth and, further, that the greatest goal in life was pleasure and one's own enjoyment. Although Charvaka did not endure as a school of thought, it influenced the development of a new way of thinking which was more grounded, pragmatic, and eventually encouraged the adoption of empirical and scientific observation and method.

Cities also expanded during this time and the increased urbanization and wealth attracted the attention of Cyrus II (the Great, r. c. 550-530 BCE) of the

Persian Achaemenid Empire (c. 550-330 BCE) who invaded India in 530 BCE and initiated a campaign of conquest in the region. Ten years later, under the reign of his son, Darius I (the Great, r. 522-486 BCE), northern India was firmly under Persian control (the regions corresponding to Afghanistan and Pakistan today) and the inhabitants of that area subject to Persian laws and customs. One consequence of this, possibly, was an assimilation of Persian and Indian religious beliefs which some scholars point to as an explanation for further religious and cultural reforms.

WHO THE GODS REALLY ARE

Now that we have discussed the background in which the science and technology referred to in our ancient texts existed in Ancient Times, let us now delve into a more mysterious aspect of our mythology i.e the innumerable references in our Ancient Scriptures to Devas or Gods with whom the Ancient Man was constantly in contact with and who regularly guided him and helped in times of need.

There are two types of Gods that were worshipped by the Ancient man. The first type are the Gods of nature like the Earth (dharti Mata), Wind (Pawan Devta), Sun (Surya Devta) and Krishna as the 'Entity of the Universe'. I will be further elaborating on the second type of Gods later, presently I shall continue the discussion about Krishna being the Entity of the Universe the Universe.

To take the view of Lord Krishna being the entity of the Universe further, Chapter 11 of the Shreemat Bhagwat Geeta English translation by ISKCON (International Society for Krishna Consciousness), the Chapter 11 in which Lord Krishna shows His "Viraat Avatar" to Arjuna on the battle field, the Chapter itself is titled "The UNIVERSAL Form'.

To give you some quotations from Chapter 11 to bring the point further:-

You are the origin without beginning, middle or end. You have numberless arms, and the sun and moon are among Your great unlimited eyes. By Your own radiance You are heating this entire universe.

• ALTHOUGH YOU ARE ONE, YOU ARE SPREAD THROUGHOUT THE SKY AND THE PLANETS AND ALL SPACE BETWEEN. O GREAT ONE, AS I BEHOLD THIS TERRIBLE FORM, I SEE THAT ALL THE PLANETARY SYSTEMS ARE PERPLEXED.

• The Blessed Lord said: My dear Arjuna, happily do I show you this UNIVERSAL FORM WITHIN THE MATERIAL WORLD by My internal potency. No one before you has ever seen this unlimited and glaringly effulgent form."

In the second paragraph above, you can see that Arjuna describes what He sees as Lord's Viraat Avataar as "You are spread throughout the sky, the planets and all space between". Now, what do we call the area that covers the planets as well as the 'space between" isn't this the essential definition of the Universe?

The Lord is the Universe itself, and we are nothing but small cells on the body of the Universe. We are all a part of His body and that is why we say Lord is OMNIPRESENT, WITHIN US and OUTSIDE US. The way we can't say that we love our right hand more than our left hand similarly, He does not love and one person more than the other because we are all part of His own body.

There are two popular misconceptions about Lord Krishna's leelas, firstly, it is well known to anyone who has a keen interest in mythology that all of Lord Krishna's 16,000 wives were all captives of a demon called Narkasura and after releasing them from captivity, Lord Krishna had married them all on their request, since after their release from captivity of Narkasura, the society was not ready to accept them.

Second popular misconception about Lord Krishna is about His Raas Leelas. Now, as already discussed earlier, the Universe is based on sound, every individual within this Universe, every human being also has a sound or sur of this own and the one's who are out of synchronicity with the sound of the Universe are very aptly called "A-Sur" meaning ones without the "Sound" or the ones out of tune with the Universe. So what Lord Krishna was doing in the Raas Leela was that He was playing the sound of the Universe with His Bansari or flute and as a result of the creation of this cosmic sound, the Gopis who were all Rishis in their previous lives used to hear the sound of the Universe and the sound of their souls and used to come rushing to Lord Krishna. Those who could not make it because of being locked up in their houses used to come to meet Lord Krishna in their Astral Body Form.

Now read the last paragraph that has been quoted from the Shreemat Bhagwat Geeta above, "My dear Arjuna, happily do I show you this UNIVERSAL FORM WITHIN THE MATERIAL WORLD" now what does He mean by saying that this is His form "Within the material world"?. What he means by this is that within this material world or Maya He can be perceived as the Universe itself, however, beyond this material world He is "Devine White Light".

In His encounter with Maharaj Muchukand whom Lord Krishna woke up from his sleep of hundreds of years in order to get Kaal Yavana killed by Maharaj Muchukand. After Kaal Yavana is killed Lord Krishna shows His Devine White Light Form to Maharaj Muchukand, which is actually His form beyond the realms of Maya.

This very Devine White Light is worshipped as "Nirakaar Paar Brahm Parmeshwar" as well. It is this very Devine White Light that is referred to as "Shiva" in our Ancient Scriptures. We often mix up Lord Shiva with Lord Shankara which is a cause of major confusion in Hinduism, Lord Shiva is formless whereas Lord Shankara is the Saakar form or the form with body.

SECOND TYPE OF GODS WORSHIPPED IN HINDUISM

To understand who the Second type of Gods that are worshipped in Hinduism are, we need to analyse a few things about the Gods that we worship. Lets see what we really indisputably know about the Gods that we worship.

Who are these various entities referred to in our Ancient Texts as:-

Yaksha,

Gandharva,

Naag,

Kinnar,

Apsaras,

Lord Indra

Now, lets think about it, when our thousands of years old religious texts tell us that Gods are non-humans who live somewhere in skies, descend from the skies for a specific purpose, i.e to help mankind with advanced technologies and go back to their abode in the skies when their purpose is fulfilled, whom are they talking about?

As rational human beings and putting together the knowledge that we have gained today through our sciences, the answer to the first question itself should tell us who the Second type of Gods are.

Yes, they are Extra-terrestrial beings who help mankind in every way they can in order to make sure that mankind fulfils its ultimate destiny of becoming a peace loving society like them and humanity can bring out the divine element that is hidden within them.

The most interesting part of this information is the fact that our Ancient texts very openly and clearly tell us about these beings as not non-humans who used to go back to their lokas or star systems. It is only our present day mental conditioning that stops us from seeing this clear fact.

The various entities referred to in our Ancient Texts as:-

Yaksha,

Gandharva,

Naag,

Kinnar,

Apsaras,

Lord Indra

are all beings from other planets and galaxies who used to visit earth regularly and engage with the inhabitants of the planet.

Another example of the writing on the wall clearly spelled out in our religious texts, is the story of Kaal Yavan's father Rishi Sheshirayan. To put it very concisely the story goes like this, Rishi Sheshirayan married an 'APSARA' from 'another world' and together they had a child named Kaal Yavan, when Kaal Yavan was very young his mother had to leave him in the custody of his father and GO BACK to her "Loka" because she "could not stay on the prithvi Loka for too long" (the atmosphere was not conducive to her existence beyond a certain period).

The system of Maha Manavtantra in Hinduism, when it says that 6 months of humans (Uttarayana and Dakshinayana) is equal to one day of Gods, they are actually referring to the way the time passes on these other galaxies.

Again we can refer to the story of Maharaj Muchukand who went to the Swarga Loka/Galaxy to help the Gods in a fight, he fought there for a period equivalent to 1 year of Swarga Loka and by the time the war was over 100 years had passed on earth and his entire empire and loved ones had moved on. Thus showing that there are galaxies where time passes more slowly than here on earth.

There are several references in our religious texts about various ExtraTerrestrial species like Gandharva- Who are CLEARLY referred to as "celestial beings" in all our religious texts, how much more proof do we want of this.

The various species of ETs who were present on Earth at that time are repeatedly mentioned in our religious texts as Yaksh, Gandharva, Naag, Kinnar.

We also have further evidence of these ET's having children on Earth. Greek hero Hercules was the son of 'God' Zeus and human woman Alcmene, meaning thereby that he was born to a human mother and ET father.

There is an entire series on history Channel called the 'Ancient Aliens' where they shows proofs from around the world that humanity has been visited by ETs since time immemorial.

These Extra-terrestrials are so clearly and openly mentioned in our scriptures that we even have Ancient stone carvings and reliefs clearly depicting Nagas changing shape from Snake to Human.

At Mahabalipuram, Tamil Nadu, we have the world's second largest relief "Arjuna's Penance" (picture on the next page) a stone carving which clearly depicts a snake coming out from the ground and changing shape into a human.

If you look at the relief closely the central of the relief clearly depicts a snake coming out from the ground which is in its original form in the beginning and changes form to human as it comes up.

ARJUNA'S PENANCE

In fact, the Ellora Caves in India there is an entire hidden network of tunnels used by the Nagas which has still not been discovered by the Archaeologists. If you visit the Ellora caves, in Maharashtra, you will find several air shafts on the floor of the caves to allow oxygen to reach underground and several underground tunnel opening all around the caves which are lying in neglect without anyone bothering to investigate them.

For anyone, who still has doubts that Hinduism has anything to do with the Extra-Terrestrials, Dr Steven M Greer; a country doctor in America; (who i personally believe is an Ancient Rishi reborn) has established an organization called CSETI (Centre for Study of ExtraTerrestrial Intelligence) where he teaches the ANCIENT VEDIC PRACTICES to contact ETs to the common public.

Dr Greer says that after reading the Vedas, he has found a way to contact the ETs and get them to visit the planet with their Crafts. He often takes out groups of people to open lands where he performs a Puja and then through yogic methods he gets his astral body to float into the Universe, locate the ET crafts and requests them to visit our planet. Dr Greer has succeeded in calling the ET spacecrafts innumerable times. This man has been working tirelessly to allow humanity to realise its potential. He has also in 2013 released one of the biggest public funded movies on ETs called 'Sirius'.

THE LAW OF KARMA

There are various galaxies and star systems in the Universe which have circumstances far better than what we face on this planet earth, there are planets which have no diseases, no wars, as per our scriptures no one grows old in "Dev Lokas". There are however, certain laws of the Universe applicable to all souls entering these galaxies or planets, the laws are called the laws of "Karma" the souls in their previous lives should have conducted themselves in a certain manner to qualify to be born on these better planets. Our religions are here to guide us on these ways to conduct ourselves on this planet so that we can qualify as advanced souls and be born in better circumstances on other planets hereafter. I will further elaborate on this in the next Chapter.

The ultimate goal for humanity is to get a moksha and get a peaceful society. The day we are able to achieve this goal, Moksha is the end of the death and rebirth cycle and is classed as the fourth and ultimate artha (goal). It is the transcendence of all arthas. It is achieved by overcoming ignorance and desires. It is a paradox in the sense that overcoming desires also includes overcoming the desire for moksha itself.

There are three ways to salvation are:

(1.) The karma-marga (the path of duty) or the dispassionate discharge of ritual and social obligations.

(2.) The jnana-marga (the path of knowledge) which is the use of meditation with concentration preceded by a long and systematic ethical and contemplative training through yoga to gain insight into one's identity with brahman.

(3.) The bhakti-marga (the path of devotion), adherence to a personal god.

The Vedas have been provided to humans for this very purpose, remember the Vedas are described as "Apurushaya" means 'not of human origin', since they have been given to us by these Extra Terrestrials to help us evolve.

THE ACTUAL SECRET OF THE JOURNEY OF THE SOULS

Vedic description of the Universe tells us about 7 lokas or 7 Star Systems.

These Lokas or Star systems are:

- Satya Loka- Abode of Brahma
- Tapa Loka-Abode of the Ascetics
- Jana Loka
- Mahar Loka
- Swar Loka
- Bhuvar Loka
- Bhu Loka.

What we actually mean by the journey of the soul is the laws of the Universe which allow a certain type of souls to reside in a certain type of a Galaxy based on its previous Karma. Now this is the reason why all the religions around the world pay so much emphasis on regulating our behaviour on this planet, because once our life on this planet ends our souls go over to the other planets in other galaxies according to their Karma.

This is the valuable knowledge being imparted to us for the benefit of mankind in the form of the knowledge of the journey of the soul in all religions today.

For those critics who would say that there is no soul, I have this to say. All the ancient civilizations around the world tell us stories like the story of Noah from the Bible and the story of Manu from Hinduism which tell us about various flood myths, about how there were cataclysmic floods in the past which swept away entire human civilizations and the survivors of the floods had to start civilization from a scratch. Till date historians refuse to believe these stories dismissing them as myths, however after the discovery of Dwarka city 100 feet under water as stated previously, these myths have been proven to be true. The reason for dismissing these stories as myths was that we did not have the equipment to search the floors of the ocean 100 ft under water at that time when these stories were declared to be myths in early 20th century when present day modern history was being written and therefore due to lack of technology, since these theories could not be proven, they were dismissed and a wide gap was created between mythology and history which has not been filled even today. But today when we have the technology to search the depths of the ocean, the truth has come out by itself.

Similarly, all the religions around the world talk about the existence of the soul and ask us to live our lives in a manner that would be beneficial for the soul, but the modern sciences yet again dismiss the idea of a soul, saying that there is no proof of it. The fact of the matter is that we don't have the requisite technology to prove it yet, but that does not mean that the soul does not exist. However, we cannot wait for the modern sciences to catch up on this information because humans are living their lives every day and need to follow the rules for the soul in order to attain a better afterlife.

Now, the question arises, that if ETs are here to benefit humanity then why don't they come forth and reveal themselves the way they did in Ancient times. The answer to this question that i can understand is that if we jog our memory back, up to the time of Lord Christ, the occurrence of miracles was quite common. After the time of Jesus these open interactions stopped. The fact that Lord Jesus himself was born through Virgin Mary without any male contribution was easily believed by the general public of the time shows that these miracles were common place, in addition to this there are various reference to voice coming from the sky and telling people that Jesus was a favoured Son of God. However, after the time of Lord Jesus, these open miracles stopped and we do not have much record thereafter of any miracles occurring openly where God descended on earth or there were any voices talking from the heaven. Therefore, in my understanding, these open interactions with ETs stopped due to the way we treated Lord Jesus Christ.

Therefore, from this we can see that though we still have access and guidance from up above which comes in our dreams/through telepathy etc., we still cannot interact freely with them in the physical world because of the inhumane treatment given by us to Lord Jesus despite the knowledge of the fact that he was here to help us and to help humanity reach its ultimate goal and yet the way we treated him lead them to be vary of us. This is not to say that Lord Jesus was an ET but only to say that when we treat someone who comes with a message of love so badly, what would anyone looking at us think. Even today if we see an ET aircraft, the first thing that we will do is probably to shoot it down. Therefore, as man has grown powerful over a period of time he has also grown violent and this is the reason why they only interact directly with people like Dr Steven Greer mentioned above who convey a clear message of peace for humanity.

Now think about the message that Lord Jesus Christ brought with him, he said we should share; "if you have two shirts, give one to the man who has none". Humans decided not to follow his advice, now think about it, what was the attitude that was the root cause of colonialism and ultimately the two world wars, to sum it up in one line, the attitude was that 'if i have 2 shirts and the other person has one i want his shirt as well, i don't care if he has none left to wear thereafter'. If we had followed Lord Jesus Christ's advice when he gave it, there would have been no colonialism and hence World War-I and World War-II could also have been avoided. It would have saved humanity a lot of grief, but all as we chose to ignore his advice.

These messengers of God know the direction in which humanity is headed and therefore, always come down on this planet to guide us in the right direction.

THE GREAT EMPIRES OF ANCIENT INDIA

Persia held dominance in northern India until the conquest of Alexander the Great in 330 BCE who marched on India after Persia had fallen. Again, foreign influences were brought to bear on the region giving rise to the Greco-Buddhist culture which impacted all areas of culture in northern India from art to religion to dress. Statues and reliefs from this period depict Buddha, and other figures, as distinctly Hellenic in dress and pose (known as the Gandhara School of Art). Following Alexander's departure from India, the Mauryan Empire (322-185 BCE) rose under the reign of Chandragupta Maurya (r. c. 321-297 BCE) until, by the end of the third century BCE, it ruled over almost all of northern India.

Chandragupta's son, Bindusara (r. 298-272 BCE) extended the empire throughout almost the whole of India. His son was Ashoka the Great (r. 268-232 BCE) under whose rule the empire flourished at its height. Eight years into his reign, Ashoka conquered the eastern city-state of Kalinga which resulted in a death toll numbering over 100,000. Shocked at the destruction and death, Ashoka embraced the teachings of the Buddha and embarked on a systematic program advocating Buddhist thought and principles.

He established many monasteries, gave lavishly to Buddhist communities, and is said to have erected 84,000 stupas across the land to honor the Buddha. In 249 BCE, on pilgrimage to sites associated with the Buddha's life, he formally established the village of Lumbini as Buddha's birthplace, erecting a pillar there, and commissioned the creation of his famous Edicts of Ashoka to encourage Buddhist thought and values. Prior to Ashoka's reign, Buddhism was a small sect struggling to gain adherents. After Ashoka sent missionaries to foreign countries carrying the Buddhist vision, the small sect began to grow into the major religion it is today.

The Mauryan Empire declined and fell after Ashoka's death and the country splintered into many small kingdoms and empires (such as the

Kushan Empire) in what has come to be called the Middle Period. This era saw the increase of trade with Rome (which had begun c. 130 BCE) following Augustus Caesar's incorporation of Egypt into the newly established Roman Empire in 30 BCE. Rome now became India's primary partner in trade as the Romans also had already annexed much of Mesopotamia. This was a time of individual and cultural development in the various kingdoms which finally flourished in what is considered the Golden Age of India under the reign of the Gupta Empire (320-550 CE).

The Gupta Empire is thought to have been founded by one Sri Gupta (`Sri` means `Lord') who probably ruled between 240-280 CE. As Sri Gupta is thought to have been of the Vaishya (merchant) class, his rise to power in defiance of the caste system is unprecedented. He laid the foundation for the government which would so stabilize India that virtually every aspect of culture reached its height under the reign of the Guptas. Philosophy, literature, science, mathematics, architecture, astronomy, technology, art, engineering, religion, and astronomy, among other fields, all flourished during this period, resulting in some of the greatest of human achievements.

The Puranas of Vyasa were compiled during this period and the famous caves of Ajanta and Ellora, with their elaborate carvings and vaulted rooms, were also begun. Kalidasa the poet and playwright wrote his masterpiece Shakuntala and the Kamasutra was also written, or compiled from earlier works, by Vatsyayana. Varahamihira explored astronomy at the same time as Aryabhatta, the mathematician, made his own discoveries in the field and also recognized the importance of the concept of zero, which he is credited with inventing. As the founder of the Gupta Empire defied orthodox Hindu thought, it is not surprising that the Gupta rulers advocated and propagated Buddhism as the national belief and this is the reason for the plentitude of Buddhist works of art, as opposed to Hindu, at sites such as Ajanta and Ellora.

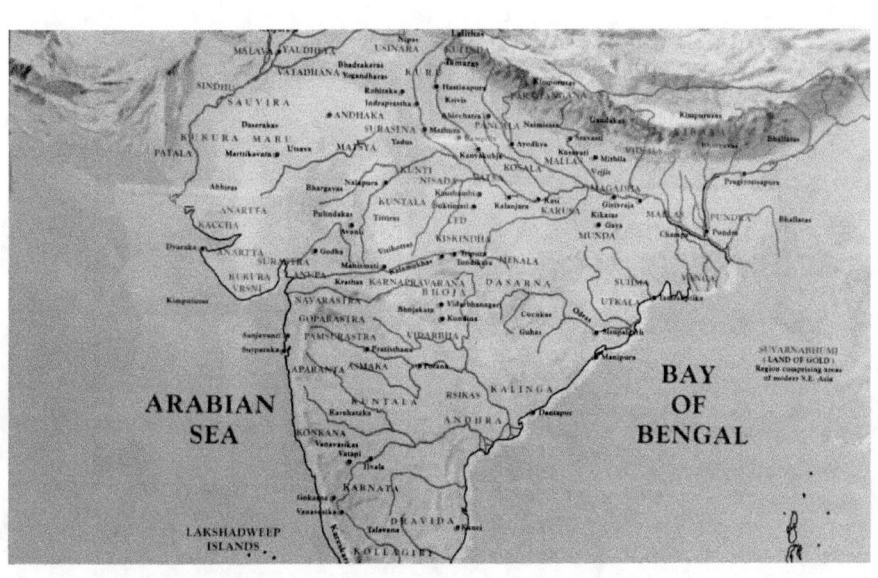

THE DECLINE OF EMPIRE & THE COMING OF ISLAM IN INDIA

The empire declined slowly under a succession of weak rulers until it collapsed around 550 CE. The Gupta Empire was then replaced by the rule of Harshavardhan (590-647 CE) who ruled the region for 42 years. A literary man of considerable accomplishments (he authored three plays in addition to other works) Harsha was a patron of the arts and a devout Buddhist who forbade the killing of animals in his kingdom but recognized the necessity to sometimes kill humans in battle.

He was a highly skilled military tactician who was only defeated in the field once in his life. Under his reign, the north of India flourished but his kingdom collapsed following his death. The invasion of the Huns had been repeatedly repelled by the Guptas and then by Harshavardhan but, with the fall of his kingdom, India fell into chaos and fragmented into small kingdoms lacking the unity necessary to fight off invading forces.

In 712 CE the Muslim general Muhammed bin Quasim conquered northern India, establishing himself in the region of modern-day Pakistan. The Muslim invasion saw an end to the indigenous empires of India and, from then on, independent city-states or communities under the control of a city would be the standard model of government. The Islamic Sultanates rose in the region of modern-day Pakistan and spread north-west.

The disparate world views of the religions which now contested each other for acceptance in the region and the diversity of languages spoken, made the unity and cultural advances, such as were seen in the time of the Guptas, difficult to reproduce. Consequently, the region was easily conquered by the Islamic Mughal Empire. India would then remain subject to various foreign influences and powers (among them the Portuguese, the French, and the British) until finally winning its independence in 1947.

GREEK CIVILIZATION

Greek Civilisation flourished in Greece more than 2000 years ago. There arose many independent city-states, which developed a remarkable system of government. The development of city-state was a unique feature of Greek civilisation. Each city was enclosed by a wall for protection. Inside the city, there was a fort called Acropolis which was situated on a hill top.

Among the Greek city-states, the most famous were Athens and Sparta. Athens was rich and cultured. Athenian citizens included writers, philosophers, artists and thinkers. The society was based on slave labor, but the citizens enjoyed a democratic form of government. You will read about Democracy in detail the later lessons. Sparta was almost like an army camp, where everyone was expected to obey the superiors. Sparta had the best army in Greece. Training in warfare was considered to be the most important thing here.

There was considerable rivalry between Athens and Sparta. But they fought side by side to drive off the mighty Persian army of Darius I and Xerxes, who tried to conquer Greece. Under Pericles, Athens enjoyed a 'Golden Age'. But a long war between Athens and Sparta, called the Peloponnesian War, which lasted for 27 years resulted in the defeat of Athens.

Do you know that Ancient Greece had the distinction of being called the birth place of Western Civilisation? They were pioneers in art and learning, science, literature and sculpture. Socrates, Plato and Aristotle were great philosophers whose works are studied even today. Herodotus and Thucydides were famous historians. Archimedes, Aristarchus and Democritus were great scientists. Aeschylus, Sophocles and Aristophanes were great dramatists. Homer was the author of the famous epics - Iliad and Odyssey.

The Greeks also had great knowledge of architecture. They built many beautiful temples and palaces. The Greeks believed in many gods. Each city had its own protector god or goddess. The gods were believed to live on Mount Olympus. The Olympic Games, first recorded in 776 BC was held every four years at a place called Olympia. Sports and athletic events were held to honor Zeus, the king of gods.

The Greek towns were the centers of administration as well as cultural and economic activities. The farmers mainly grew grapes, olives and grain. Wine and olive oil were important products. The Greeks, at one time, also established vast empires. Alexander of Macedonia, better known to history as Alexander the Great, led his army out of Europe and conquered Syria, Mesopotamia, Egypt, Afghanistan and even parts of Central Asia and North-Western India. This led to the spread of Greek ideas and learning. Alexander died at an age of thirty-three only. After his death, his empire broke up into smaller kingdoms. Later, Greece was conquered by the Romans.

ROMAN CIVILIZATION

In 510 BC, the Romans set up a Republic on the city of Rome which is on river Tiber in Central Italy. The Roman Republic was ruled by the senate, which consisted of a group of elders called senators. They elected two Consuls each year to lead them. By 200 BC, Rome became the leading power of Italy. It was able to defeat rivals like Carthage for the control of the Mediterranean world.

In the early Roman society, there were three classes of people – the patricians (aristocrats), the plebeians (commoners) and the slaves. Roman economy was based on slave labor. Rich Romans owned slaves. These slaves were often trained for the gladiators' fight, which was a fight between the slaves and wild animals. There were also frequent slave revolts in Rome. One such revolt was led by Spartacus in 74 BC.

Although Rome was a Republic, strong and influential leaders fought for power. Julius Caesar was one such leader who got enormous power and became a dictator. In 44 BC, Caesar was murdered and a civil war broke out. After the war, Augustus Caesar became the first emperor of Rome. The Roman Empire spread to three continents – Europe, Asia and Africa. Do you know that it was during the rule of Augustus, the great prophet, Jesus Christ appeared? He was the founder of Christianity. He was born in Bethlehem. According to him, all men and women are the children of God. He taught people to love each other. After his death, the followers of Christ spread his teachings among the people. At its peak, the Roman Empire stretched from Mesopotamia in the east to Gaul and Britain in the west. People throughout the Empire adopted Roman way of living. Towns with baths, temples, palaces and theatres were built. In the countryside, the Romans built huge, comfortable farmhouses called villas. Roman rulers used to preside over victory parades, religious ceremonies and games in the arenas and amphitheatres. Gladiator's fight, chariot racing, and theatre were some of the common amusements.

The Roman Empire was divided into several provinces, each ruled by a governor. He had a number of officers under him who looked after different affairs of administration. The main fighting forces of the Roman army were the legions. Each legion had 5000 soldiers headed by a commander. The Roman Empire was governed by the personal will of the emperor. But his power depended on the army. Weak emperors were often overthrown by the army generals.

By 395 AD, the huge Roman Empire was divided into two halves for better governance. The Eastern part with capital at Byzantium survived even after the fall of the Western Roman Empire in the face of barbarian invasion in 476 AD. Emperor Constance gave Byzantium a new name – Constantinople. It became the home of Eastern Orthodox Christian faith and the capital city of the Byzantine emperors.

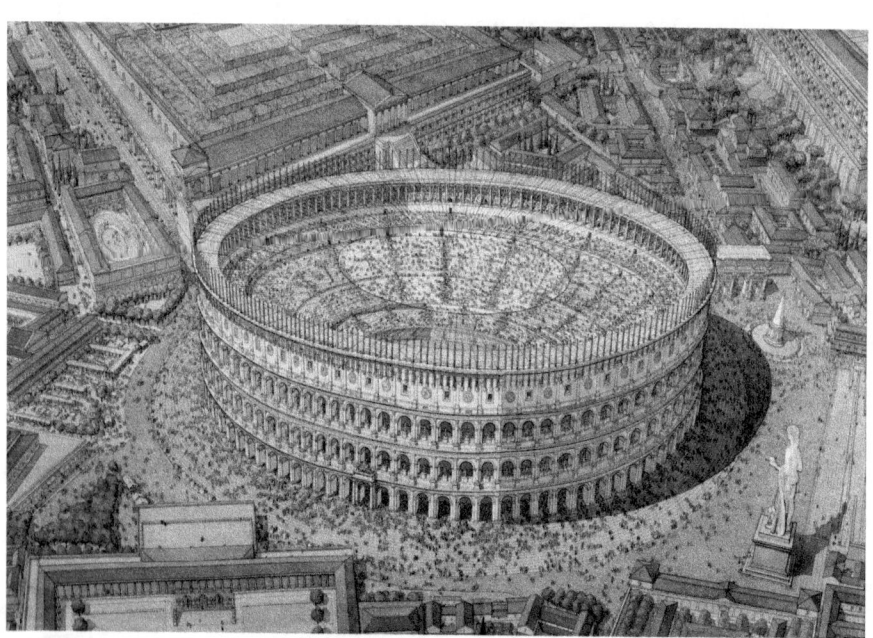

PERSIAN CIVILIZATION

In the Iron Age, Persia (Modern Iran) was inhabited by the Aryan communities. One branch of them, known as the Medes, settled in the western part of the country. Another branch occupied the southern and eastern parts and was called the Persians. The Medes built up a powerful kingdom covering a vast area of Iran. At first, the Persians also had to acknowledge the supremacy of the Medes. One of the Persian kings, Cyrus, united the Persians in 550 BC. He built a powerful army and successfully conquered Babylon, Assyria and Asia Minor. Darius I was the greatest emperor of Persia. He belonged to the Achaemenian dynasty. His empire stretched from River Indus to the Eastern shores of Mediterranean Sea. He made Persepolis his capital in 518 BC. During his reign, Persian art, architecture and sculpture flourished. He also built a powerful navy.

The Persian emperors were able administrators. They divided their empire into provinces, each governed by a Satrap or Governor. The Persians were good soldiers with strong cavalry, navy and had iron weapons. Though the Persians were defeated by Alexander the Great in 331 BC, their civilisation did not come to an end. Their culture and civilisation continued to flourish under the Parthian and Sassanian emperors. But ultimately they were conquered by the Arabs in 7th century AD.

Like the Indo-Aryans, the early Persians worshipped the forces of nature. They believed in the sun god, sky god and other gods. They considered fire to be a symbol of holiness. They also performed fire rites and practiced animal sacrifices. Later, a religious preacher Zoroaster found the religion called Zoroastrianism. He taught them about Ahura-Mazda, the Lord of Heaven and Light, who gives men strength and energy. According to Zoroaster, life was a constant struggle between good (light) and evil (darkness). The holy scripture of the Persians is called Zend – Avesta.

BUDDHISM

Gautama Buddha, the founder of Buddhism, was born in 563 BC at Lumbini which is situated near the Indo-Nepal Border. He was the son of Shuddhodhana, the chief of Shakya of Kapilavastu. At the age of 29, Gautama left home and attained Bodhi (enlightenment) at Bodhgaya under a pipal tree. He delivered his first sermon at Sarnath near Varanasi. His teachings included four Noble Truths (Arya Satya) and Eightfold Path (Ashtangika Marga).

According to Buddha:

(i) the world is full of misery (dukkha);

(ii) desire (trishna) is the cause of this misery;

(iii) if desire is conquered, then all sorrows can be removed;

(iv) this can be done by following the Eight fold Path; which included:

(a) right memory (b) right aim (c) right speech

(d) right action (e) right livelihood (f) right efforts

(g) right memory and (h) right meditation

Buddha suggested a 'Middle Path' - away from both extreme luxury as well as extreme austerity. He also laid down a code of conduct such as non-killing and nonstealing for his followers. He died at the age of 80 (483 BC) at Kushinagar in Uttar Pradesh.

JAINISM

Rishabhanath, the first Tirthankara, is known to be the founder of Jainism. Vardhamana Mahavira was the 24th Tirthankara of this sect, Parshvanath being the 23rd one. Mahavira was born in 540 BC at Kundagram near Vaishali (Bihar). His father was the Chief of Jhatrika Kshatriya clan. Mahavira became an ascetic at the age of 30 years and died at Pawapuri in 468 BC near Rajagriha. His followers came to be known as 'Jainas'.

Jainism had no place for a supreme creator. It recognized the existence of gods, but placed them lower than the Jaina teachers. The main aim of Jainism is the attainment of freedom from worldly bonds. Like Buddhism, Jainism opposed the ritualistic practices and evils of Vedic Brahmanism. It also opposed the caste system and accepted the doctrine of Karma and rebirth. Jainism has five cardinal principles:

(i) Ahimsa or non-violence,

(ii) Truthfulness,

(iii) Abstention from stealing,

(iv) Non-attachment, and

(v) Celibacy or Brahmacharya. The three jewels (Triratna) of Jainism are:

(a) Right vision (Samyak Darshana),

(b) Right knowledge (Samyak Jnana), and

(c) Right Conduct (Samyak Charita)

CHINESE CIVILISATION

The Chinese civilisation grew up in the Hwang Ho valley in North China. The first rulers known were the Shangs (1523 BC to 1122 BC), who built China's first cities. They also contributed to art and culture. The Chinese writing system was developed during this period. Craftspersons of this period, especially the bronze workers were great experts in their fields.

The Shang dynasty was overthrown by the Zhous, who built strong forts and walled towns to defend themselves from invaders. It was during the later phase of the Zhou rule that iron was introduced, thus ending the Bronze Age in China.

In 221 BC, the Chin rulers came to power in China. They ordered the use of common language, common laws and common weights and measures throughout their empire. Do you know that they were the rulers who built the famous Great Wall of China?

After the Chins, the Han dynasty came to power, who ruled till AD 220. It was during this period that Chinese traders had contact with the West through the famous Silk Route, crossing Central Asia and Persia. The people of China worshipped a number of deities. Worship of ancestors, nature and spirits were very common. In China a famous religious preacher named Confucius advocated a system of right behavior, which greatly influenced Chinese society and government. He laid emphasis on good moral character, respect to elders and loyalty to the family and obedience to the laws of the State.

MOHENJO-DARO & HARAPPAN CIVILIZATION

The Indus Valley Civilization dates to c. 7000 BCE and grew steadily throughout the lower Gangetic Valley region southwards and northwards to Malwa. The cities of this period were larger than contemporary settlements in other countries, were situated according to cardinal points, and were built of mud bricks, often kiln-fired. Houses were constructed with a large courtyard opening from the front door, a kitchen/workroom for the preparation of food, and smaller bedrooms.

Family activities seem to have centered on the front of the house, particularly the courtyard and, in this, are similar to what has been inferred from sites in Rome, Egypt, Greece, and Mesopotamia. The buildings and homes of the Indus Valley peoples, however, were far more advanced technologically with many featuring flush toilets and "wind catchers" (possibly first developed in ancient Persia) on the rooftops which provided air conditioning. The sewer and drainage systems of the cities excavated thus far are more advanced than those of Rome at its height.

The most famous sites of this period are the great cities of Mohenjo-Daro and Harappa both located in present-day Pakistan (Mohenjo-daro in the Sindh province and Harappa in Punjab) which was part of India until the 1947 partition of the country which created the separate nation. Harappa has given its name to the Harappan Civilization (another name for the Indus Valley Civilization) which is usually divided into Early, Middle, and Mature periods corresponding roughly to 5000-4000 BCE (Early), 4000-2900 BCE (Middle), and 2900-1900 BCE (Mature). Harappa dates from the Middle period (c. 3000 BCE) while Mohenjo-Daro was built in the Mature period (c. 2600 BCE).

Harappa's buildings were severely damaged and the site compromised in the 19th century when British workers carried away a significant amount of material for use as ballast in constructing the railroad. Prior to this time, many buildings had already been dismantled by citizens of the local village of Harappa (which gives the site its name) for use in their own projects. It is therefore now difficult to determine the historical significance of Harappa save that it is clear it was once a significant Bronze Age community with a population of as many as 30,000 people.

Mohenjo-Daro, on the other hand, is much better preserved as it lay mostly buried until 1922. The name Mohenjo-Daro means `mound of the dead` in Sindhi and was applied to the site by local people who found bones of humans and animals there, as well as ancient ceramics and other artifacts, emerging from the soil periodically. The original name of the city is unknown although various possibilities have been suggested by finds in the region, among them, the Dravidian name `Kukkutarma', the city of the cock, a possible allusion to the site now known as Mohenjo-Daro as a center of ritual cock-fighting or, perhaps, as a breeding center for cocks.

Mohenjo-Daro was an elaborately constructed city with streets laid out evenly at right angles and a sophisticated drainage system. The Great Bath, a central structure at the site, was heated and seems to have been a focal point for the community. The citizens were skilled in the use of metals such as copper, bronze, lead, and tin (as evidenced by artworks such as the bronze statue of the Dancing Girl and by individual seals) and cultivated barley, wheat, peas, sesame, and cotton. Trade was an important source of commerce and it is thought that ancient Mesopotamian texts which mention Magan and Meluhha refer to India generally or, perhaps, Mohenjo-Daro specifically. Artifacts from the Indus Valley region have been found at sites in Mesopotamia though their precise point of origin in India is not always clear.

DECLINE OF HARAPPAN CIVILIZATION

The people of the Harappan Civilization worshipped many gods and engaged in ritual worship. Statues of various deities (such as Indra, the god of storm and war) have been found at many sites and, chief among them, terracotta pieces depicting the Shakti (the Mother Goddess) suggesting a popular, common worship of the feminine principle. In c. 2000 - c.1500 BCE it is thought another race, known as the Aryans, migrated into India through the Khyber Pass and assimilated into the existing culture, bringing their gods and the language of Sanskrit with them which they then introduced to the region's existing belief system. Who the Aryans were and what effect they had on the indigenous people continues to be debated but it is generally acknowledged that, at about the same time as their arrival, the Harappan culture began to decline.

Scholars cite climate change as one possible reason noting evidence of both drought and flood in the region. The Indus River is thought to have begun flooding the region more regularly (as evidenced by approximately 30 feet or 9 meters of silt at Mohenjo-Daro) and this destroyed crops and encouraged famine. It is also thought the path of the monsoon, relied upon for watering the crops, could have changed and people left the cities in the north for lands in the south. Another possibility is loss of trade relations with Mesopotamia and Egypt, their two most vital partners in commerce, as both of those regions were undergoing domestic conflicts at this same time.

Racialist writers and political philosophers of the early 20th century, following the lead of the German philologist Max Muller (l. 1823-1900), claimed the Indus Valley Civilization fell to an invasion of light-skinned Aryans but this theoy has now long been discredited. Equally untenable is the theory that the people were driven south by extra-terrestrials. Among the most mysterious aspects of Mohenjo-daro is the vitrification of parts of the site as though it had been exposed to intense heat which melted the brick and stone. This same phenomenon has been observed at sites such as

Traprain Law in Scotland and attributed to the results of warfare. Speculation regarding the destruction of the city by some kind of ancient atomic blast, however, (the work of aliens from other planets) is not generally regarded as credible.

PARALLEL UNIVERSE & TIME

The Big Bang Theory is the leading explanation of how the universe began. Some scientists believe that when this happened, an alternate universe was created? They believe it is an anti universe, a mirror image of our own, where even time runs backwards. People there find things which are normal to us, weird, and which are weird to us, normal. It may be because they have always experienced things the other way!

Our Universe is infinitely humongous, yet there are theories of the existence of more such universes in the world. There are so many myths and truths that revolve around the science of multiverse.

According to Hindu mythology, the creator of this universe, Brahma had pleaded Lord Vishnu to come down to Earth to control the despotic dominance of tyrants. That is when Lord Vishnu incarnated as Krishna unbeknownst to all the Lords and Gods. Lord Krishna grew up in the village of Vrindavan from where he amended the world into a happy place. Brahma, who had no clue about Krishna's true identity, was bewildered by Krishna's triumph over all the despots. That is when he decided to find who Krishna was. Assuming it to be the mischief of an invincible wizard, he tried to coax him into an unkind prank. One fine day, when Lord Krishna and his friends went to the woods for a picnic, Brahma vanished all of his friends. Krishna, who finds out that it was Brahma who was behind this deed, replicates himself as his friends to avoid disturbing the routine of his village. He kept the parents happy by taking the place of their kids. After a year, when Brahma returned to Vrindavan to see how the villagers were responding to his trickery, he was bewildered on seeing all the kids playing around Vrindavan. Shocked, Brahma goes to Lord Krishna.

However, to his astonishment, he sees thousands of Brahmas flying towards him on swans. Lord Krishna then unfolds the truth of parallel universes by telling Brahma that he is the creator of only this universe. He explains to him that there are many powerful creators and many more enormous universes than his'.

There are fictitious ideas saying that parallel universes do exist, exactly like our universe. It is just that their wars might have had different outcomes than the ones we know. Species that are extinct in our universe might have evolved and adapted in others and there, we humans might have gone extinct.

Though this is the fictional side, scientifically, there was a thought experiment that was comprehended by Albert Einstein. It is called the Schrodinger cat experiment where a cat and a pile of unstable gunpowder was kept inside a box. As the name emphasizes, it was carried out by the Austrian physicist Erwin Schrodinger. The probability of gunpowder exploding is equal to the probability of gunpowder doing nothing. According to our viewpoint, when we look into the box, the cat is either dead or alive. The experiment is repeated enough times until the odds of the cat staying alive and losing its life are equal.

This was our interpretation of the cat's life but when we consider the cat's perspective, there are two possibilities where the gunpowder explodes and the cat sees it exploding or it doesn't explode and the cat stays alive. So the cat either survives or dies. This means that if the cat lives, the other possibility of it losing its life is dissolved by nature. Here, it is us who are looking at the cat's fate of life but, if we look further and imagine ourselves in the cat's place, there might be someone else observing. There might be different outcomes for every act of ours. So, there can be a chance that all the possibilities are happening in a larger multiverse with many parallel universes, just like how we are seeing the cat live or die every single time we run the experiment. Probably, there is another you who might have chosen to read a different one!

Some people also believe that deja vu (the feeling of having lived through the present moment already) is something experienced by your 'universal twin' in a parallel universe that is ahead of time.

A few real-life experiences also amplify the existence of parallel universes. In 2015, the people of China saw a towering city of skyscrapers that appeared in the clouds. Some people believe that a parallel universe opened up for a few minutes while others call it a mirage.

In 1954, there was another man who landed at the Tokyo airport with a genuine passport of a country called Taured. But Taured, according to our world map, doesn't exist! He had authentic proofs belonging to a bank, hotel and the company he was working in. But neither the hotel nor the bank nor the company have any records of him. The next day, he puzzlingly disappeared from the room he had checked into.

In current time scientifically, there are not many legitimate proofs to make sure that parallel universes exist. Although there are some mythological stories and some real-life instances, and in the end, it is an individual's choice to believe in it or not.

According to sanatan-dharma parallel universe

What some scientists have recently discovered and accepted, has been stated in the scriptures of Sanatan-Dharma for more than 5000 years. In the cosmos, there are many universes and they are spiritually linked with each other (parallel universes) to form a multiverse.

Lord Krishna's pastimes are eternally ongoing in the material creation. This means all His transcendental activities that He performed 5000 years ago here on earth are continuously performed in other universes. For example, He spoke the Bhagavad-Gita to Arjuna 5000 years ago here on earth. He is speaking the Bhagavad-Gita to Arjuna right now in another universe, and once He is finished, He will start to speak it to the same Arjuna in another universe and so on... All His activities are continuously repeated in different universes, one after the other.

"The consecutive pastimes of Krsna are being manifested in one of the innumerable universes moment after moment. There is no possibility of counting the universes, but in any case some pastime of the Lord is being manifested at every moment in one universe or another." (Sri Catanya-caritamrta Madhya-Lila 20.382)

"Since all Krsna's pastimes are taking place continuously, at every moment some pastime is existing in one universe or another. Consequently, these pastimes are called eternal by the Vedas and

Puranas." (Sri Catanya-caritamrta Madhya-Lila 20.395)

Arjuna is with Lord Krishna in multiple universes at the same time, to hear the Bhagavad-Gita and take part in other pastimes with Him. In a similar way, it is possible that some of us can also be present in other universes right now and/or we could repeat our life again on earth or in other universes. Time is eternal, we (the soul) are eternal, and the universes are unlimited. Thus, the rules of probability imply that events or history will be repeated.

What is Time and why it is Eternal?

"The Supreme Personality of Godhead said: Time I am..." (Lord Krishna, Bhagavad-Gita 11.32) "I am also inexhaustible time..." (Lord Krishna, Bhagavad-Gita 10.33)

Time is one of the energies of Lord Krishna. Since Lord Krishna is eternal, His energies like time are also eternal.

Before creation, God and His energies existed

Something cannot come from nothing. Everything comes from Lord Krishna OR Lord Narayan (God).

"Brahma, it is I, the Personality of Godhead, who was existing before the creation, when there was nothing but Myself." (Lord Krishna, Srimad-Bhagavatam 2.9.33)

BODY CHAKRA

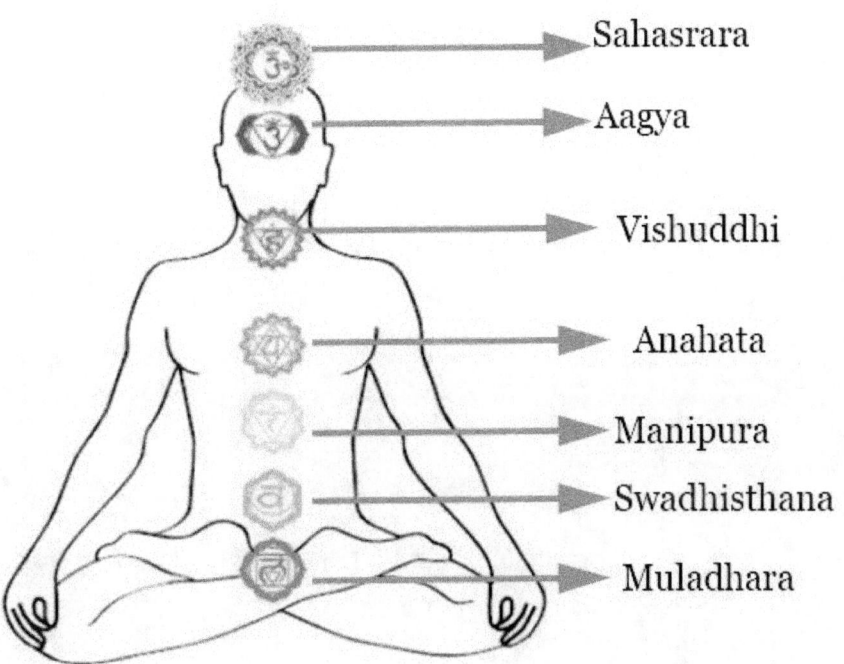

ACCORDING TO SANATAN-DHARMA (HINDU DHARMA)

The Sanatan-Dharma scriptures do not support the theory of a big bang creating the universe. The universes are created by Maha-Visnu, an expansion of Lord Krishna.

Lord Krishna expands as Maha-Visnu and with every exhalation, millions of universes come out from the pores on His body. With inhalation, they all go back into His body. The lifespan of our universe is just one breath of Maha-Visnu, 311.040 trillion years.

The Creation and Universes Are Expanding

"Arjuna could see in the universal form of the Lord the unlimited expansions of the universe..." (Bhagavad-Gita 11.13)

"O son of Kuntī, at the end of the millennium all material manifestations enter into My nature, and at the beginning of another millennium, by My potency, I create them again." (Lord Krishna, Bhagavad-Gita 9.7)

Millennium in this verse refers to the lifespan of the universe, 311.040 trillion years. Currently the universe is 155.522 trillion years old, and after another 155.518 trillion years, the universe will be destroyed and then after some time, He will create again. The scientists estimate the universe to be at least 13.7 billion years old, and this number keeps increasing as the scientists develop better technologies to measure. Other religions estimate the universe to be around 6,000 years old.

The universes and their engineers (Brahmas), appear from the pores of Maha-Visnu and remain alive during His exhalation. (Reference: Sri Brahma-samhita 5.48)

"Combining all the different elements, the Supreme Lord created all the universes. Those universes are unlimited in number; there is no possibility of counting them. The first form of Lord Viṣṇu is called Maha-Viṣṇu. He is the original creator of the total material energy. The innumerable universes emanate from the pores of His body. These universes are understood to be floating in the air that Maha-Viṣṇu exhales. They are like atomic particles that float in sunshine and pass through the holes of a screen. All these universes are thus created by the exhalation of Maha-Viṣṇu, and when Maha-Viṣṇu inhales, they re-enter His body. The unlimited opulences of Maha-Viṣṇu are completely beyond material conception." (Sri Caitanya-caritamrta Madhya-Lila 20.277-280)

The Creation and Annihilation are Cyclic

"The whole cosmic order is under Me. Under My will it is automatically manifested again and again, and under My will it is annihilated at the end." (Lord Krishna, Bhagavad-Gita 9.8)

The Big Bang Theory

The big bang theory is that a small, dense object exploded and our universe was created. It has been expanding since the bang.

The scientists cannot explain the origin of the dense object nor the bang.

Law of Conservation of Energy

The law of conservation of energy states that energy cannot be created nor destroyed, but it can change from one form to another or transferred from one object to another. The total energy of the universe or a closed system is a constant.

If energy can't be created or destroyed how did the dense object from which the universe was created come into existence in the universe? The scientists have no answer.

The scientists are unable to scientifically prove the theory of big bang. They cannot create big objects by banging a tiny object. Thus, the big bang theory is just a theory without clear evidence.

What is the Origin of the Energy in the Universe?

"My Self is the very source of creation..." (Lord Krishna, Bhagavad-Gita 9.5)

The energies come from Lord Krishna (God). Time and matter are energies of Lord Krishna. That's why time and matter is eternal, because God is eternal. His energies are also eternal.

Before the creation, matter existed in another form as energy of Lord Krishna. At creation, the matter changed from one form to another, basically what we see today, solid stones, planets, metal, and so on. This is also accepted by the scientists as the Law of Conservation. Energy cannot be created nor destroyed but it can change forms.

Matter and Living Entities are Eternal

"Material nature and the living entities should be understood to be beginningless. Their transformations and the modes of matter are products of material nature." (Lord Krishna, Bhagavad-Gita 13.20)

The above verse states that matter (material nature) and the souls (living entities) are eternal. This is because material nature is the energy of God and the souls are His parts. God is eternal and so are His energies and parts.

"This material nature, which is one of My energies, is working under My direction, O son of Kunti, producing all moving and nonmoving beings. Under its rule this manifestation is created and annihilated again and again." (Lord Krishna, Bhagavad-Gita 9.10)

Lord Krishna is in full control of the creation and annihilation of millions of universes.

We (the soul) are the God particle

"The living entities in this conditioned world are My eternal fragmental parts. Due to conditioned life, they are struggling very hard with the six senses, which include the mind." (Lord Krishna, Bhagavad-Gita 15.7)

All living entities (souls) are God particles or sparks of God.

THE SCIENCE OF TRAVELLING FASTER THAN THE SPEED OF LIGHT

One can travel faster than the speed of light, but only by using a spiritual method. With mystic yoga perfections, one can travel light years in a moment simply by thinking of the destination.

"The ten secondary mystic perfections arising from the modes of nature are the powers of freeing oneself from hunger and thirst and other bodily disturbances, hearing and seeing things far away, moving the body at the speed of the mind, assuming any form one desires, entering the bodies of others, dying when one desires, witnessing the pastimes between the demigods and the celestial girls called Apsaras, completely executing one's determination and giving orders whose fulfillment is unimpeded."
(Lord Krishna, Srimad-Bhagavatam 11.15.6-7)

Lord Krishna is everywhere in every atom in every universe and we are His particles. By meditating on Him, one can achieve the eighteen mystic yoga perfections, which includes the ability to travel at the speed of the mind, appear at multiple locations at the same time, and inter universal travel. These perfections have been achieved and practiced by millions of yogis, including Narade Muni, who spends his life travelling from one universe to another.

"If one thinks of the Supreme Personality of Godhead and quits his body, he will certainly reach the spiritual planets..." (Lord Krishna, Bhagavad-Gita 8.13)

If one thinks of Krishna at the time of death, he (the soul) will instantly be transferred many light years away to the spiritual manifestation.

"Engage your mind always in thinking of Me, become My devotee, offer obeisances to Me and worship Me. Being completely absorbed in Me,

surely you will come to Me." (Lord Krishna, Bhagavad-Gita 9.34)

If a pure devotee wishes, then he (the soul) can instantly be transferred to be with Lord Krishna in another universe where He is currently performing His transcendental pastimes.

(References: Srimad-Bhagavatam canto 11, chapter 15: Lord Krsna's Description of Mystic Yoga Perfections)

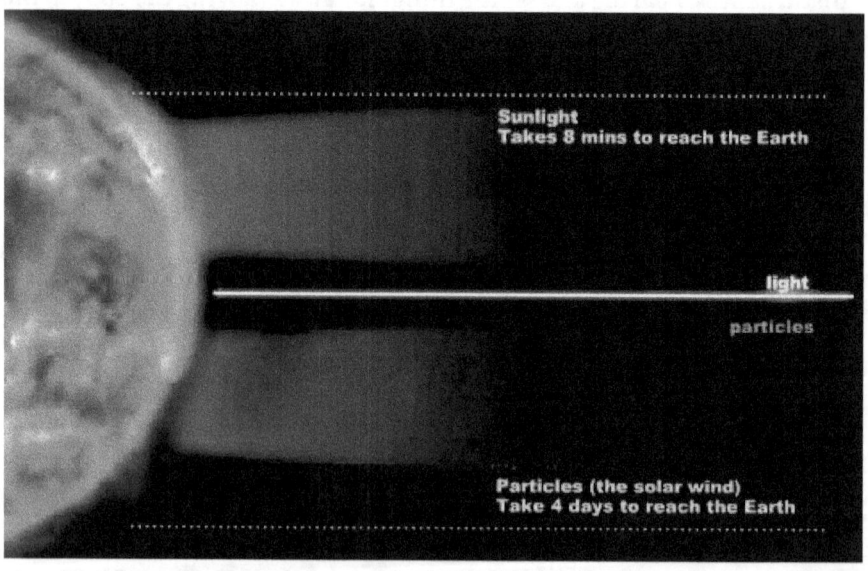

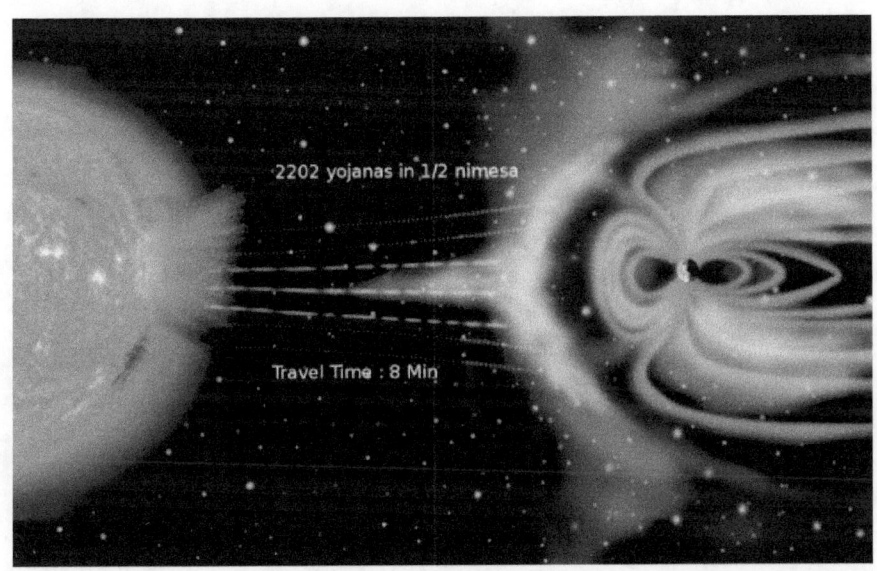

THE SCIENCE OF CREATION OF LIVING BEINGS

Based on the Sanatan-Dharma knowledge, there is no evolution of species. All the species of living beings in the universe exist from the moment of creation. (Reference: Bhagavad-Gita 9.8 purport)

The population of humans in the world originate from thousands of couples and not just one as taught by some religions.

"By human calculation, a thousand ages taken together form the duration of Brahma's one day. And such also is the duration of his night. At the beginning of Brahma's day, all living entities become manifest from the unmanifest state, and thereafter, when the night falls, they are merged into the unmanifest again. Again and again, when Brahma's day arrives, all living entities come into being, and with the arrival of Brahma's night they are helplessly annihilated." (Lord Krishna, Bhagavad-Gita 8.17-19)

One daytime of Brahma is 4.32 billion years. During this time, thousands of human couples are created and we originate from them.

ACCORDING TO SANATAN-DHARMA THE SOUL OF YOU

Anyone can see that the material bodies of living beings (humans, plants, and animals) are physically different, temporary, situated in different conditions, circumstances, and locations. Within each body of living beings, there is the soul, which are all equal and eternal.

- You are the soul and not the body
- All living beings (humans, plants, and animals) have a soul
- All souls are equal, one soul is neither inferior nor superior to another
- The soul is not Brahmin, nor Ksatriya, nor Vaisya, nor Sudra
- The soul is not high caste, nor low caste
- All souls are eternal, full of knowledge, and completely blissful
- Every soul has an eternal companion relationship with God
- Every soul is a family member of God
- Every soul is accompanied by the Super Soul (God) in the heart. He is the witness, permitter, and the well-wisher
- Every soul gets liberation eventually, there is no eternal damnation
- The soul cannot be destroyed by any weapon, burned, blown, nor moistened
- The soul is neither male nor female
- The soul is neither black nor white
- The soul is not Indian, nor American, nor African

- The soul is not Hindu, nor Christian, nor Muslim
- The soul is neither diseased nor healthy
- One soul is neither richer nor poorer than another
- The soul is not old, nor young, and does not age
- The soul has no weight and cannot be seen with the eye
- The soul is situated in the heart of all living beings
- The size of every soul is 1/10,000th of the tip of the hair

The Law of Conservation of Energy Supports the Bhagavad-Gita Science of the Soul Knowledge

Energy cannot be created nor destroyed but it can be transferred from one object to another. We (the soul) are the energy and the body is the form. When our current form (the body) is destroyed, we move into another form (another body).

"That which pervades the entire body you should know to be indestructible. No one is able to destroy that imperishable soul." (Lord Krishna, Bhagavad-Gita 2.17)

Number of Species of Living Beings

"jalaja nava lakshani, sthavara laksha-vimshati, krimayo rudra-sankhyakah, pakshinam dashalakshanam, trinshal-lakshani pashavah, chatur lakshani manavah" (Padma Purana)

- Jalaja (Water based life forms) - 0.9 million
- Sthavara (plants and trees) - 2.0 million
- Krimayo (Reptiles) - 1.1 million
- Pakshinam (Birds) - 1.0 million
- Pashavah (animals) - 3.0 million
- Manavah (human-like) - 0.4 million

Total 8.4 million species of living beings.

A soul occupies a Body exactly based on its Desires and Deserves

The living entity (soul) is placed in a particular body, universe, planet, country, City, street, house, room, and the womb of a particular mother - exactly according to his past deeds. Life is an intelligent design, nothing happens by chance.

"The living entity in the material world carries his different conceptions of life from one body to another, as the air carries aromas. Thus, he takes one kind of body and again quits it to take another." (Lord Krishna, Bhagavad-Gita 15.8)

"The living entity in material nature thus follows the ways of life, enjoying the three modes of nature. This is due to his association with that material nature. Thus, he meets with good and evil among various species." (Lord Krishna, Bhagavad-Gita 13.22)

We are the Eternal Soul and not the current temporary Body

Energy (the soul) cannot be destroyed, but it can be transferred from one body to another

"Never was there a time when I did not exist, nor you, nor all these kings; nor in the future shall any of us cease to be. As the embodied soul continuously passes, in this body, from boyhood to youth to old age, the soul similarly passes into another body at death. A sober person is not bewildered by such a change." (Lord Krishna, Bhagavad-Gita 2.12-2.13)

"The soul can never be cut to pieces by any weapon, nor burned by fire, nor moistened by water, nor withered by the wind." (Lord Krishna, Bhagavad-Gita 2.23)

When a person dies, we say 'he passed away', but the body is right in front of us. This implies there are 2 entities, the 'body' and the 'he'. The 'he' is the soul, the real person. This is the first teaching from the Bhagavad-Gita, the real 'he' is the soul and not the body. The body is temporary but the 'he' is eternal. Where has he (the soul) passed away to? Another body.

Death is not the End, it's the start of a new Life

"One who has taken his birth is sure to die, and after death one is sure to take birth again. Therefore, in the unavoidable discharge of your duty, you should not lament." (Lord Krishna, Bhagavad-Gita 2.27)

"As a person puts on new garments, giving up old ones, the soul similarly accepts new material bodies, giving up the old and useless ones." (Lord Krishna, BhagavadGita 2.22)

THE SCIENCE OF SENSE GRATIFICATION

"While contemplating the objects of the senses, a person develops attachment for them, and from such attachment lust develops, and from lust anger arises. From anger, complete delusion arises, and from delusion bewilderment of memory. When memory is bewildered, intelligence is lost, and when intelligence is lost one falls down again into the material pool." (Lord Krishna, Bhagavad-Gita 2.62-2.63)

"As a strong wind sweeps away a boat on the water, even one of the roaming senses on which the mind focuses can carry away a man's intelligence." (Lord Krishna, Bhagavad-Gita 2.67)

"When one dies in the mode of passion, he takes birth among those engaged in fruitive activities; and when one dies in the mode of ignorance, he takes birth in the animal kingdom." (Lord Krishna, Bhagavad-Gita 14.15)

"Life's desires should never be directed toward sense gratification. One should desire only a healthy life, or self-preservation, since a human being is meant for inquiry about the Absolute Truth. Nothing else should be the goal of one's works." (Srimad-Bhagavatam 1.2.10)

- Object orientation develops attachment
- Attachment develops lust
- Lust develops anger
- Anger develops delusions
- Delusions lead to bewilderment of memory
- Memory bewilderment leads to loss of intelligence
- Loss of intelligence leads to ignorance
- Ignorance leads to degradation and lower species

वो देवेषु नव यज्ञा गर्भिः । तनूह द्वे विप्रमतिश्चकारेवे तं कल्याण वसु विश्वमिव वा
मंत्रेप्रमतिरिव तासिन खवैयस्व तव नीम यो वर्यो सत्तारायः शुतिनः संसहस्रिणः ।
सुवीरीयं तित्रुतपामदाप्य ॥ ३३ ॥ त्वामग्रेप्रथममा युमायै वेदेवा अकृण्वन्नदुहस्यवि
र्पति । इ ॐ मकृण्वन्मनुषस्य शसनी पितुर्य तुत्रो ममकस्य ताय तं नो अग्ने वेदेते
पाकृजिमे चोनोरक्षतन्नेश्ववेद्य । मातातो कस्यते नेचुण वास्यनिम चंरश्रमाण ह तवे
न्ते मंत्रुय्ये वप फुरंतरो निषुगायचतुराध्वद्यासोयोगान ह्योवेकाय धायसकी
रश्चिय्ये उमनसावेनाविता लेमन्त्र उरुशसायवा घतेस्या हैप हेकृण्णः परमंव नोपित
नो । आ आर्त्स्ये तुतधर्मनिरच्यसे पिनाष्रपाकं शासिश्रि दशाधिदुष्षरः । लेमंत्रे प्रयनदक्षि

SPIRITUALITY ACCORDING TO THE CULTURE

A sense of freedom is something that we all want to achieve. Being independent in whatever aspect it may be definitely boosts our confidence and makes us do better in life. We must all be equipped with this certain type of positive energy within us in order to have a major shift in the way we live.

Spiritual Empowerment

Spirituality may be associated with religious things and ceremonies but in this case, it does not necessarily mean that we should be hooked to a religion. Experiencing this state would mean that one's consciousness is awakened. This enables the person to see the person one really is and become aware of the capabilities and limitations attached to it. This makes the person become happy and contented with the person that he is. Thus, he is able to take care of and understand himself more than he used to. Being spiritually empowered makes a person aware of what makes him happy and makes him more sensitive to what would make other people happy.

Why Is It Important

Our society today has embedded in us stereotypes and perfect models as to how a person should be. This makes most of us disgruntled and embarrassed about our own selves. Some people have even gone to worse conditions, sulked, get into depression, and later on, even get suicidal tendencies.

However, when a person becomes spiritually empowered, he sees himself for who he is and who is not. He becomes aware of his capabilities and thus he knows what actions are to be done. Empowered people know their roles in society and they know what they can do to bring about change in their selves, in others, and in the environment.

If we were just able to empower ourselves spiritually, then freedom, in whatever aspect of life, would just be within our reach.

Spiritual Empowerment through Alternate Therapies

One can reach spiritual empowerment in many ways. The most common of these methods are through alternative therapies. :-

Yoga : Yoga uses the mental as well as physical disciplines that originated in India. Most Yoga practitioners today use yoga as a form of exercise. However, this form of meditation may also be used to gain Moksha.

Moksha is the state where a person attains liberation and freedom from all worldly sufferings. It comes from a Sanskrit word that literally means release or to let go. At the end, the person is able to find his own identity called the Supreme Brahman. Doing yoga may also help a person have a stable relationship with himself while experiencing calm and peace.

According to the Bhagavad Gita, the three paths to moksha are

- **karma-marga**
- **jnana-marga**
- **bhakti-marga.**

Reiki : This method originated in Japan and means spiritual power from a Chinese loanword. Reiki originated with Mikao Usui after a 21-day retreat in Mount Kurama. Practitioners of Reiki aim to abide to its principles that include (in a translation):-

- Do not be angry
- Do not worry
- Be grateful
- Work with integrity
- Be kind to others

Reiki also makes use of a universal spiritual energy that could actually have a healing effect. Anyone can also gain of this energy but has to go through a process of attunement done by a Reiki master.

Er Mei Qi Gong

This type of practice believes that a unique form of matter called the Qi can be transmitted to others to provide healing and promote good health as well as help in spiritual empowerment. Some people even believe that they are able to develop their skills in clairvoyance and telepathy.

The choice of alternative therapy, if at all, would really depend upon you and in which practices you are most comfortable of doing. In addition, there are many other therapies that you could undergo aside from the ones listed here.

Being Your Own Spiritual Coach

It is not only enough to seek help from other people by using meditative and alternative therapies on your way to spiritual empowerment. It would help much if you were able to become the personal coach to your own self so that you can criticize easily if you make any mistakes.

Having an enriched and empowered spiritual life would mean that a person is able to accept one's self, no matter the limitations he has. When you fail to accept that and tend to blame yourself for things because you are a weak person who has too little capabilities, then you should start to tutor yourself into forgetting that part. Keep in mind that you are in the road to spiritual empowerment, if you do not help yourself improve and accept things as they are, nobody else can help you in that aspect.

It is also important that you always remember the mistakes you have done in the past and take action from it instead of regressing and going back the other way. Do not forget that mistakes are what shapes a person and sometimes, they can be inevitable. One can always learn something from a mistake and when he does, he commits never to make the same action again.

However, you must also remember that a person learns and improves. You cannot always say that it is unavoidable to make a mistake because it can be – when it happens repeatedly. Making the same mistakes over again is a sign that you are not doing anything to help yourself improve and move a notch higher in terms of spiritual empowerment. Become your own coach – you are the only person who can always accompany yourself and check for any mistakes that you might commit. You are also the only person who can bring yourself not ever to commit the same mistake again.

Learning from life and moving on

When you go through the process of spiritual development, it is not always a guaranteed success at the start. There are many times that a person will fail but it is in getting back up that he gets courage and determination to withstand whatever situation he has to face in the future.

If you are on this journey, it is very important to have the will to achieve your goal – of being independent, free, and empowered as well. It takes a little hard work to achieve this as well as some patience. Even though the path you will be taking will have some bumps ahead, you must bear in mind that everything will end soon enough and if you focus your mind to it, you should not notice that you have already reached your goal.

It is then important that each time you fail and make a mistake; you learn to accept that you also have some limitations in you. You cannot be perfect but you are trying to be the best that you can be. When you make mistakes, which are often inevitable at the start, it is normal to feel sad and a little depressed. However, this failure should not be a reason for you to sulk and even regress and go back to your old ways. Instead, you should use it as a stepping-stone so that you can move on further to your spiritual goal.

When you become successful in this journey, it should feel very pleasant and light. You have reached your goal because of all your hard work and your determination. Courage is also important so that you can bravely face your shortcomings and learn to accept it. Flaws are a part of who a person is and it cannot be avoided that he has one however, it should not be a hindrance as to why he could not move on to his goal.

Spirituality and Money Understanding the Equation

When we try to achieve spiritual empowerment, it can be hard to go against all the temptations in the world. It can be hard to avoid indulging ourselves with what money has to offer.

As they say, money makes the world go round and many times, we would forget that there are other forms of happiness in the world that no amount of money can buy. However, this is not the case for spiritual people.

When One is Spiritual

A spiritual person is more aware of himself or herself. The person is more aware of his surroundings and to the sufferings of the people. This is the reason why many of them are not after material richness. These types of people find more happiness and fulfillment in the little joys of life that money is not able to bring. Thus, they are able to have a different view and concept about money, what it can do to our lives and to the people surrounding us as well. These type of people minds is steady. they are beyond from happiness & saddness. There life is steady. These type of people really good in focusing anything, fearless they never fear with death and they will live lustless life. They can achieve more than that like sidhi or nidhi if they focus too much on god.

- **Their Views on Money:** The use of money, of course, cannot be avoided. However, a person who has gained spiritual enlightenment only sees money as a means to an end and not as anything else. Those who want to do something to society work hard not for themselves but so that they could provide the needs of the people around them. They give out to charities and help other people in need without expecting to gain anything out of it. Enlightened people do not chase after money. Instead, they accept it as it comes. Money is still needed to buy essentials but not for the things that could bring happiness. These people believe that they could get what they want out of money but it does not mean that all their happiness relies in it.

Understanding the Secret of the Law of Attraction

Did you ever know that spiritual empowerment had something to do with Quantum Physics? You may not believe it but it does. Here is why:-

A Quantum Physics Law

According to experts, the Law of Attraction traces its roots to Quantum Physics. However, many of those in the scientific community believe that this law already belongs to pseudoscience.

The Law of Attraction states that one can use his energy to his own advantage when he follows four principles:-

- You must have specifics of what you want to achieve.
- You must ask the world to give it to you.
- You must think and feel that you have already gotten what you really want to have.
- You must be prepared for its coming and let go of whatever connections that comes with it.

Thus, the law states that when you think of something as already happening to you, there is a big probability that it will truly happen to you.

How It Is Connected to Spiritual Empowerment

You already know that spiritual empowerment and enrichment is what you want to achieve. When you start to think of it as something that is already in you, you also start to act as if you already have that empowerment. Thus, your actions will help you work out what you need to do as things follow. Later on, you will never notice, that you have already reached what you want to achieve and you let go of the other things that is stopping you from achieving it.

When you are in your journey for spiritual enlightenment, it is important that if you feel and start to act the way an enlightened person should, you would also keep out any negative vibes. Because what you think would attract the energy around you, it is best to keep out negative thoughts from your mind to keep out negative energy as well.

THINKING & GROWING RICH

The Law of Attraction has been found with many uses, including in the aspect of becoming rich. According to this law, your thoughts have an energy that can attract other circumstances in life. Although this has been dismissed as a part of pseudoscience, there are still many instances in our lives where we believe that the Law of Attraction takes place.

On Becoming Rich

The Law of Attraction says that if a person thinks of a specific goal in mind, feels, and knows as if it is already a truth, then things will start to follow and the universe will seem to conspire to make it a reality. Many people believe that when they start to do the same in their want to become rich, they also become one. Of course, this event does not happen instantly. When you start to think that you are already rich, or are becoming one, your actions start to modify as well. You unconsciously do things that could actually make you rich and succeed in it which is why so many of us think that the whole world willed for it to become true when the truth is, it was actually more because of our own doings.

On Succeeding in the Endeavor

If you want to succeed, this should also be something you want to keep in mind. A person who believes in success would have self-confidence that could take him places. When a business associate or a client meets up with someone who possesses this quality, they also become confident in making a deal with the person because they can trust that person and they would believe that they could also bring them success. In its simplest terms, the Law of Attraction works this way. A person who has positive energy is sure to attract positive energy as well and block out the negatives, guaranteeing success.

STRIKING THE BALANCE BETWEEN THE INNER & OUTER ASPECT OF YOUR BODY

Being spiritually empowered brings us not only to awareness of ourselves but also of others and our surroundings as well. However, there are times when these concepts could clash, which is a reason why balance must be maintained between the two.

The Inner Aspects of the Body

An enlightened person becomes more fully aware of his own self, his body, his capabilities as well as the limitations and flaws. However, a person who is spiritually empowered is able to accept whatever weaknesses they have, work on it and make it a stepping stone for improvement and not as a hindrance.

The Outer Aspects of the Body

While there are needs in the body that has to be fulfilled, we must also remember the others that we have to think of as well. You may have experienced freedom and independence for yourself but this is useless when you see around you that the people are not having the same experience. This is a time when the principles that you wish to live by begin to clash.

Striking a Balance between the Two

Usually, people who have experienced this freedom would also want to impart in others what they too have experienced. It is not enough to be able to feel happiness within one's self but in the interaction with other people as well. This would ensure that person that he is not only working things for himself but also to his surroundings, thus creating a balance.

In addition, happiness felt within oneself is incomplete if the person finds that his surroundings could not give him the happiness and enlightenment that he had worked so hard to achieve. It is for this reason that these people also try to spread and teach what they know so that a balance between the happiness of the outside and inside aspects of the body is achieved.

MOVING CLOSER TOWARDS SPIRITUAL NIRVANA

Every journey has an end and each undertaking you get yourself into has a goal to be achieved at the end. As you go through the process of spiritual empowerment, you do not only get yourself to experience freedom and independence but you move close to nirvana as well.

What is Nirvana?

Nirvana is said to come from a Pali word meaning, "blowing out". Thus, this means that a person experiencing nirvana has blown out greed and hatred and is free from suffering. In Buddhism, Nirvana is said to be a state where a person achieves and experiences perfect peace of his own mind and frees himself from cravings, anger, and other afflictions. He also becomes at peace with the world, gives kindness and consideration to other people, and does not obsess with physical things anymore.

How Is It Achieved?

According to the Pali Canon, nirvana can be achieved in many ways. First, it can be achieved from insight and self-awareness alone or it can be achieved through understanding. Nirvana may also be achieved through the deeds and righteousness that a person has as well as the virtues, understanding, and consciousness. It can also be achieved with some effort and concentration or through the four foundations of mindfulness, which include the body, sensations, mind, and mental contents.

If you summarize all of these things, the path to spiritual empowerment, which ultimately leads to nirvana, is the same as the Threefold Training of Buddha, which included wisdom, mental development, and virtue. A person who is able to lead a life with the right speech, action, livelihood, effort, mindfulness, concentration, understanding, and intention is sure to be able to reach the right path to enlightenment. When this happens, the person is not anymore attached to the material world and has found a different kind of happiness and contentment in their lives.

Conclusion

Empowering the spirit is a very tall order. It requires high discipline, a particular frame of mind and much sacrifice. But when you attain the state, there's not much you would wish for.

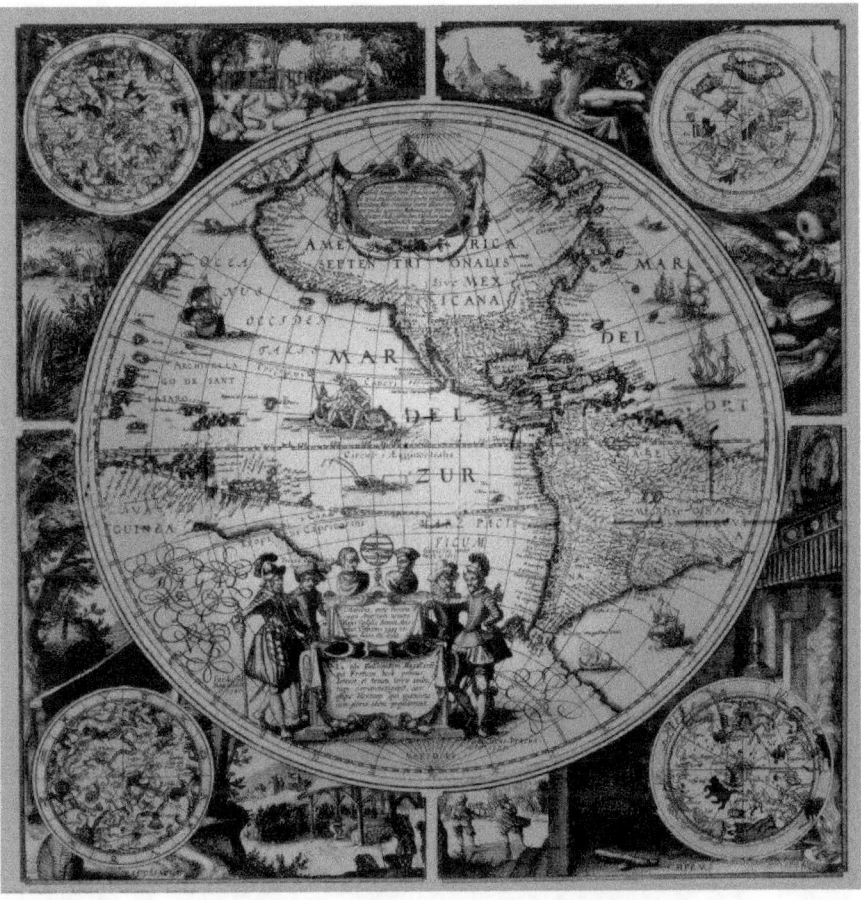

••••

So, this is the INTER GALACTIC ANCIENT, whom we have been trying to understand since thousands of years.

FEEDBACK AND RESPONSE

Author's email id: Sandeepbisht256@gmail.com

Contact no.: +919319908754

Please provide your views

www.ingramcontent.com/pod-product-compliance
Lightning Source LLC
Chambersburg PA
CBHW070643220526
45466CB00001B/271